T0180992

IT Enabled Services

Shiro Uesugi
Editor

IT Enabled Services

 Springer

Editor
Shiro Uesugi
Graduate School of Business Administration
Matsuyama University
Ehime
Japan

ISBN 978-3-7091-1688-3   ISBN 978-3-7091-1425-4 (eBook)
DOI 10.1007/978-3-7091-1425-4
Springer Wien Heidelberg New York Dordrecht London

Printed on acid-free paper

Springer is part of Springer Science+Business Media (www.springer.com)

Preface

The focus of this book is IT-Enabled Services. It has been nearly a quarter of a century since the commercialization of the Internet. Informatization of the Service Industry has occurred rapidly during the last half of the previous century and it has completely transformed production processes in general. Furthermore, the development of the Internet has been transforming the process and the nature of the service. Together with the trend toward a more service-oriented economy, new services are being born and new ways of business are being created.

The objective of this book is to provide an integrated volume of conceptual, theoretical, and practical cases that, together, offer some exciting perspectives on this fascinating area of research. The idea of creating this book emerged from the research presented, and ensuing debate in a series of workshops of IT-Enabled Services. We began reviewing the service aspects of the Internet in our first ITeS Workshop at the IEEE/IPSJ Symposium on Applications and the Internet (SAINT) 2005 and eventually included wider aspects of Information Technology of services. This publication is to share our interests with a wider range of readership.

Bearing this in mind, the structure of this book is designed to serve as a textbook, a conceptual book, a case book as well as a basis for research resources. The authorship includes researchers in such diverse fields as economics, engineering, policy science, sociology, medicine, mathematics, computer science, moral philosophy, pedagogical science, and management studies. Also, the approach is global because the authors are from Australia, Germany, Japan, New Zealand, Qatar, Taiwan, Thailand.

The contents of this book are largely divided into two parts, namely, theoretical precepts and topical issues. Dr. Tetsuro Kobayashi (National Institute of Informatics) and Dr. Hitoshi Okada (National Institute of Informatics) provide a theoretical and evidence-based article about the trust mechanism of buyers in electronic commerce. Dr. Yuya Dan (Matsuyama University and Ludwig-Maximilians-Universität München) is invited to describe a mathematical analysis on diffusion in social networks. Professor Kiyoshi Murata (Meiji University) is invited to provide an important managerial perspective on IT professionals.

The latter part includes topics such as trust, medical practices, government, and education, together with cases about Taiwan, Thailand, and Japan and chapters about privacy-related issues. Professor Shigeichiro Yamasaki (Kinki University) contributes a conceptual framework about trust mechanism and the architecture of IT-Enabled Services with trust. Dr. Eizen Kimura (Medical School of Ehime University) proposes the development of a new medication system using smartphones. Eltahir Kabbar and Dr. Peter Dell (Curtin Business School) provide an analysis about problems in relation to the e-Government development index. Paul Spijkerbosch (Matsuyama University) looks at research on electronic language learning.

Professor Yu-Hui Tao (National University of Kaohsiung) and Dr. C. Rosa Yeh (National Taiwan Normal University) present the practices of IT-Enabled Services in education in Taiwan. Dr. Nagul Cooharojananone (Chulalongkorn University) and Dr. Kanokwan Atchariyachanvanich (King Mongkut's Institute of Technology Ladkrabang) provide case studies of Mobile Payment and Internet Banking in Thailand. Doctors Takashi Okamoto and Nobuyuki Soga (Ehime University) present a case study about Japanese students' behavior toward electronic commerce.

The last part consists of various privacy-related issues in IT-Enabled Services. Because the service industry relies heavily on personal involvement, privacy issues are critical in this research. Professor Hirotsugu Kinoshita (Kanagawa University) describes a model-based approach about information exchange while protecting privacy. Dr. Yohko Orito (Ehime University) discusses digital identity and its related problems in the management of information privacy. Dr. Hidenobu Sai (Ehime University) offers a discussion about the social media platform in relation to Human Flesh Search.

I would like to express my sincere thanks to all participants of past workshops who contributed the cross-disciplinary debates and brought a wide variety of ideas, excellent research, efforts, and insightful visions. Matsuyama University's *Chiiki Kenkyu Center Project* (Regional Research Center Project) 2008–2010 "Research on ITeS in region" supported this research.

All authors are grateful to Springer for their cooperation and help, especially Stephen Soehnlen and Annelies Kersbergen, in putting this volume together. Also, our thanks are towards Athiappan Kumar, who completed the final type setting so to make this book coming to alive.

Our intention in this book is to offer new views which incorporate the wide range of ITeS development in the first decade of the twenty-first century. It is our wish to share our views with readers of not only academia but also business persons and policy practitioners.

Matsuyama, Japan   Shiro Uesugi

Contents

Information Technology-Enabled Services

As the global economy turns more and more service oriented, Information Technology-Enabled Services (ITeS) require greater understanding. Increasing numbers and varieties of services are provided through IT. Furthermore, IT enables the creation of new services in diverse fields previously untouched. Because of the catalyzing nature of internet technology, ITeS today has become more than "Outsourcing" of services. This book illustrates the enabling nature of ITeS with its entailment of IT, thus contributing to the betterment of humanity. The scope of this book is not only for academia but also for business persons, government practitioners and readers from daily lives. Authors from a variety of nations and regions with various backgrounds provide insightful theories, research, findings and practices in various fields such as commerce, finance, medical services, government and education. This book opens up a new horizon with the application of Internet-based practices in business, government and in daily lives. *Information Technology-Enabled Services* works as a navigator for those who sail to the new horizon of service oriented economies.

Chapter 1
IT-Enabled Services

Shiro Uesugi

1.1 Introduction

In this chapter, the concepts and theories of IT-enabled Services (ITeS) are revisited. In general, "IT" stands for "Information Technology." In the discussion of ITeS, however, "IT" should include "Communication" so that "ITeS" means "Information Communication Technology-enabled Services."

It is important to make this clarification in considering the "Service" aspects taken up in the following sections. The rapid development and penetration of IT in people's daily lives and the consequent drastic changes in society are primarily due to the advancements in the fields of IT and communications. Therefore, even though IT implies "communication," it is necessary to reaffirm this implication.

This chapter looks at previous works on the stage theory developed by Nolan in 1973. After reviewing Nolan's original theory, the work of Shimada and Takahara (1993), a Japanese scholar's development of this theory, is presented.

The review of previous work continues with a detailed study of Reponen et al. (2003a), whose book contains the term ITeS in the title, "Information Technology-Enabled Global Customer Services." This book, co-written with a team that includes Harvard Business School Professor F. Warren McFarlan, is the very first book dedicated to the study of ITeS.

It has been nearly 10 years since Reponen et al. published their work. Since then, however, there has been limited research dedicated to ITeS like Treebhoohu (2012) and Tsokota (2011). Our work since 2008 (Uesugi 2008; Dan et al. 2009; Okada and Uesugi 2009, 2010, 2011, 2012) aims to continue the efforts to study ITeS.

Second, this chapter considers the implications of technical transformation from analog to digital technology. It is necessary to review this before studying the aspects of IT relating to services. Reviews of the development of digitization and its

S. Uesugi (✉)
Graduate School of Business Administration, Matsuyama University, Ehime, Japan
e-mail: shiro.uesugi@nifty.com

S. Uesugi (ed.), *IT Enabled Services*,
DOI 10.1007/978-3-7091-1425-4_1, © Springer-Verlag Wien 2013

impact on society are presented, thus serving as the background on the relationship between digitization and services.

Third, this chapter presents a discussion of the relationship between ITeS and the production process. Why is ITeS so important? It enables the visualization of processes and consequently helps improve the quality of products. Digitization enables visualization of all kinds of processes. For consumer goods, visualization of production and sales process, such as Electronic Customer Relationship Management (eCRM) and Internet marketing, have already been developed and deployed. Through the evolution of these process visualizations, services are found to be the source of quality and competitiveness of products.

Finally, notable cases of ITeS today and their future prospects are presented. It can be safely assumed that ITeS serves as the source of improvement in products. As digitization processes and communication capacities are advanced and enhanced, respectively, the value created through ITeS increases. The potential impact of service provision in various industries, such as education, medical practice, banking, community, government, agriculture and production in general, is also presented.

1.2 Related Works

Previous works are reviewed in this section, including the stage theories of system development proposed by Nolan (1973, 1979, 1982), Shimada and Takahara (1993), and Miyakawa (1994). The studies of Reponen et al. (2003a), who are the first to focus on IT-enabled services, are also reviewed.

1.2.1 Stage Theories

Nolan developed a theory to explain the development of information systems. His first paper, which presented the concept that would later be known as stage theory, was published nearly 40 years ago. There are some sequels to the stage theory he developed along with other scholars. In the following sections, Nolan's theory is reviewed, followed by a review of the Japanese scholar's work. After reviewing the development of stage theory and its application, this chapter presents an application of the theory to the current context featuring bi-directional communication with broadband networks.

1.2.1.1 Stage Theory of Nolan

Nolan published his four-stage theory in his work entitled, "Managing the computer resource: a stage hypothesis," in *Communications of ACM* in 1973. He identified four stages in the development of information systems, namely, Stage I – *Initiation*, Stage II – *Contagion*, Stage III – *Control,* and Stage IV – *Integration*. He placed these stages on the horizontal axis and placed information-related spending on the vertical axis. Consequently, the plotted figures showed an S-shaped curve.

In 1979, Nolan revised his hypothesis to add two more stages, as described in his chapter, "Managing the Crisis in Data Processing," published in 1979. The additional ideas came in the form of two more stages, namely, Stage V – *Data Administration* and Stage VI – *Maturity*. From the fact that Nolan himself revised his hypothesis, it is apparent that the stage theory has been a popular focus of reviews in many ways, while it has continued to be a popular analytical methodology of information systems (McFarlan and Nolan 1973; Gibson and Nolan 1974; Nolan and Croson 1995). In the late twentieth century, the rapid development of the Internet has changed the structure of IT.

1.2.1.2   Diversion of Stage Theories

As a diversion from the stage theory, the work of Shimada and Takahara (1993) is reviewed in this section. The works of Shimada and Takahara (1993) and Miyakawa (1994) are two of the most popular textbooks in Management Information Systems (MIS) in Japan. Shimada and Takahara (1993) provide additional aspects to Nolan's six stage theory. They interpret Stages I to IV, which have been originally explained as "the age of Data Processing (DP)," as "the age of mainframe" in order to include historical views on the system's development. They likewise interpreted Stages IV to VI, originally referred to as "the age of Information Technology (IT)," as "the age of the personal computer (PC)." Furthermore, Shimada and Takahara reflected on the development of the Internet in the 1990s, and added "the age of the Internet" and "the age of ubiquitous net" to refer to the overlapping previous stages.

The differentiation between DP and IT reflects the development in the use of information, which has been mainly devoted to automated production processes. For example, operations research, such as those on quality control, inventory control, and optimized production planning, represent the typical application of DP. The use of information in decision making, as in MIS and Strategic Information System (SIS), is differentiated from DP and is regarded as IT.

Like the S-shaped curve illustration by Nolan's stage theory, according to Shimada and Takahara, the age of mainframe computing began in the 1950s, followed by the age of the PC (which began in the 1970s), the age of the Internet (which started in the 1990s), and the age of ubiquitous network (which started in the late 2000s). Their theory proposes that these major technologies are not mutually exclusive but have multilayered structures.

1.2.1.3   Stage of Bi-directional Communication with Broadband Network

In the twenty-first century, the development of the Internet displayed a remarkable expansion not only in geographical coverage, but also in the kind and scope of technological applications. The emergence of personal devices and the accompanying access to the Internet have both accelerated this process. In Japan, the year 2005 is regarded as the dawn of the broadband era. The broadband network environment all over the nation was almost completed, and industries and consumers were ready to connect to broadband Internet by that time.

Japan's network environment has another unique feature: the infrastructure of mobile phones. The driving force behind the diffusion of mobile phones in the country was the introduction of the i-mode service by NTT Docomo. This Internet connection service enables users to almost fully access the Internet via their mobile phones. It opened up unlimited access to the Internet via personal wireless devices in Japan, resulting in the ubiquitous Japanese Internet environment and broadband network.

The ubiquity of the Internet and the broadband network reached a new stage when netbooks, tablet terminals, and smartphones were developed and introduced by year 2010. Ubiquitous broadband became one feature of wireless mobile communication networks. From 2010 through 2011, smartphones and devices, such as the iPhone, iPad and those using the Android platform, demonstrated rapid penetration in the market. At the same time, Social Network Services (SNSs), such as Facebook and Twitter, became more popular because they can be accessed using these mobile devices.

These phenomena paved the way for the stage of bi-directional communication with broadband network. At this stage, a new situation occurred, in which "the consumers of information" became "the producers of information." This was not expected at the stage of IT. Likewise, it is different because, at this stage, anybody can become the producer of information. Thus, the concept of "the producers of information" expanded to include those who used to be considered merely as consumers.

The producers of information are those who actively upload their products to the Web. For example, even though there is a limit of 140 words per tweet, the texts and networks of re-tweets all add a huge amount of information. This is one of the outcomes of the bi-directional communication network.

1.2.2 Review of Reponen et al. (2003)

In 2003, Reponen edited and published a book entitled, *Information Technology Enabled Global Customer Service*. This 16-chapter book consists of 9 chapters of theoretical articles and 7 chapters of cases. It is the very first book to have the words "Information Technology-Enabled Service" in the title. Since this book is published, almost decade has passed. We owe greatly to works of Reponen and his colleague. It is the starting point and provides the basis for our discussion about ITeS. Therefore, it is important to review their works. The following sections present a review of the book.

1.2.2.1   The Viewpoint of Global Customer Service

The followings are the quotation from Reponen's eight-point forecast about the business environment in 2010 (Reponen 2003b, p. 5).

- Everybody would have a personal communicator that localizes, identifies, communicates, and acts as a credit card and key
- Commerce would be mainly done via networks (order, payment, supply information, etc.)
- Delivery chains would be modified so that customized physical products may be assembled close to the customer using standard parts and elements
- Services would be available 24 hours a day, seven days a week
- Most people would find work in information-related fields, and telecommuting work would increase to around 50 %
- Delivery channels would become more direct via automated networks
- Juridical companies would be founded and re-founded according to financial and legislative needs
- Labor markets would exist where work is offered through the Internet, and people accept posts, often without even knowing their employer

As it turns out, most of the forecasts stated above turned out to be accurate. In fact, the point about telecommuting and electronic commerce is particularly evident. Although the author overestimated the penetration of credit cards in the Japanese market, his predictions are mostly correct. Reponen also presents a method of analyzing ITeS based on the work of Karimi et al. (2001), who categorized industries into four types (Reponen 2003b, p. 6):

- IT-Enabled Customer Focus Firms
- IT-Leaser Firms
- IT-Lagged Firms
- IT-Enabled Operations-Focus Firms

There are two dimensions on the four categorizations as follows:

- Customer Focus
- Operation Focus [ibid.]

According to him, "a sustainable leadership position should, however, be gained with an intelligent combination of process reengineering and IT." He further added that "in IT-led firms, integration and coordination of operations will be the main challenge."

1.2.2.2   Improving the Previous Study

The works of Repponen et al. need some improvements. Their study reflected the technological environment at the turn of the century. During this time, some technologies they expected had not developed, while others they did not imagine had shown remarkable development. For example, they recognized the trend of digitization of the industry as a whole and the penetration of digital communication into the lives of individuals. However, they did not focus on "ubiquity," which is now the central concept in analyzing ICT and business models using ICT.

The prediction "everybody has a personal communicator that localizes..." (Reponen 2003b, p. 5) is correct to a certain extent, because it describes people's use of personal devices in their daily lives and in various occasions. However, the direction of the development of Japan's mobile phone market differs in many aspects.

Mobile phones, in general, are considered products that have evolved from the radio communication system. Therefore, their features are seen as extensions of the radio communication system. However, Japan's feature phones are called "Gala-kei," a term derived from the "Galapagos Syndrome" (NRI). This is because the phones now include so many service applications that are not needed in other parts of the world. Such an evolution has turned phones in Japan into as if different creatures.

More importantly, almost all the featured phones are equipped with a specific prepaid type payment system, as Reponen predicted. The IC chipset dedicated to this service is versatile enough to be used as a key, ticket, and identification device, among others. However, within the first year of the introduction of smartphones, their sales had surpassed that of "Gala-kei."

When smartphones emerged in the Japanese market, they were initially not capable of providing the services mentioned (e.g., payment) because they were not equipped with an IC chipset. Hence, users' migration from "Gala-kei" to smartphones happened quickly despite the initial inconvenience. Users sought the flexibility provided by the smartphones and the variety of software that these devices can handle. As a matter of course, the new design of the smartphones and the factor of usability, such as the touchscreen feature, attracted great attention. Yet, the deciding factor for most users is the large volume of downloadable applications. The sheer number of downloadable applications is due to the large number of production resulting from the shift in the role of users from consumers to producers, as reflected in the concepts of "Web 2.0," "Consumer Generated Model (CGM)," and "prosumers."

In other words, Reponen was unable to predict the *Kopernikanische Wende*, in that the traditional roles of "user" and "producer" became flexible. Thanks to developed infrastructure, which includes open-source architecture, broadband networks and the electronic marketplace for content, the transformation of said roles has become easier than ever.

In sum, the studies of Reponen et al. (1993a) can explain only certain aspects of today's IT-enabled Services. It is important to incorporate the perspectives mentioned above.

1.3 Review of the Transformation from Digital to Analog

It is necessary to understand that the source of the value of digitized products is the service being provided. In the competition among analog products, there are only two options to enhance competitiveness: through technological innovation and through improvements in the product's physical quality. In the case of digital products, however, this basic condition has changed. Such digital products as cameras, computers, mobile phones, television sets, and even automobiles no longer require fine tuning of physical products once the threshold of quality has been defined. In the following sections, technical differences between analog and digital products are presented, followed by a consideration of the social impact of digitization.

1.3.1 Differences Between Digital and Analog Technology

Analog technology is, in some aspects, inferior to its digital counterpart. The quality of analog products is defined by the level of physical details in contrast to digital. Digital uses the dichotomy for sampling, which determines the quality level of the finished products. Quantization of the digitization refers to the rigid definition and unmistakable presentation of what is required. Therefore, the results of digitization are completely replicable as long as they are stated and can be communicated. Analog products are bound by the limits of physical refinement capability, whereas digital products are not. Some cases of ICT are presented in the following sections.

1.3.1.1 Analog ICT

The first example worth considering is the analog calculator – a machine made of gears, which does certain actions to gain certain results. In a sense, it is similar to an astrolabe, which mimics the movement of the stars. Skilled watchmakers competed on their ability of how precisely they can copy the movements of nature. The capability of a device is a direct reflection of the skill of its producer. These devices can communicate information by acting and being recognized by the users as if they were the replicas of the original Mother Nature (Ulmann 2010).

1.3.1.2 Digital ICT

The second example is the tabulating machine invented by Herman Hollerith. In 1888, Hollerith won a competition organized by the United States Bureau of the Census, and the machine he had invented was used to process the census.

The tabulating machine operated on a simple principle. Using a hole on a paper card, the machine detected whether or not an electric current exists; if there is a hole, the machine answered "yes." This principle has been subsequently developed in the punch-card system for use in programming digital computers (Pugh 1995).

1.3.2 Digitization and Its Social Implications

As mentioned above, there are fundamental differences between analog and digital technologies. Whereas analog needs refinements of skills, digital does not. For a long time, a source of competitive advantage of the Japanese production industry is considered to be rooted in its capacity to finetune ("Suri-awase"). Originally, "Suri-awase" meant "rubbing each other," then it became "bouncing ideas off each other to come up with a precise finished product."

In the following section, the social implications of digitization are reviewed. First, the physical production process aspect is presented. Second, the production process of information is observed. Third, the development from IT to ICT and its implication is discussed.

1.3.2.1 Physical Production Process Aspect of Digitization

The Japanese "Suri-awase" process is one of the major sources of competitive power of Japanese production industries. This process requires substantial experience from skilled workers, because it demands certain levels of understanding of the tacit knowledge between production processes. Digitization, however, quantizes the processes so that even unskilled laborers can also perform the tasks involved. Hence, it is considered as a threat to Japanese competitive advantage.

This situation is characterized by the generalization of production processes. Digitization combined with the module-production system, and then enhanced by the communication system (i.e., an advanced ICT), is applied to a wide scope of physical production processes. It transforms the production process that is previously considered impossible to realize without workers' long experience, into one that is easy to replicate. This is the process of "*Mieru-ka*" or "visualization." Know-how and tacit knowledge are visualized, and the visualized knowledge is digitized and transferred to the production process. This development has also changed the way value is added on to the product. In this sense, the sources of values are now revealed as part of a visualization process.

1.3.2.2 Role of Information in the Production Process

Under the digitized production environment, two changes have been observed. One is the replacement of the production process from manual to automated, and in this mode, robots are programmed to copy the production methodology once it is digitized. For example, skilled craftsmanship has been replaced with Numerical Controlled (NC) lathe, machining centers, and factory automations.

The other change is the reduction of inventories. As the information about the demands of parts is visualized and shared between suppliers and users, extra inventories are thus eliminated. Consequently, extra production capacities are eliminated, and the level of competitiveness is increased. For example, the evolution of the analog "Kanban system" has prevailed across countries. The system originally used paper slips eventually replaced by a digitized supply chain management system, thus making it possible to conduct cross-border control of procurements, inventories, and production.

1.3.2.3 Meaning of Evolution from IT to ICT in the Production Process

IT has brought about a significant change in the production industry; this process has been accelerated by digitization. It has transformed the meaning of production from something focused on physical (tangible) goods to one that is centered on information (intangible) processing. In the first decade of the twenty-first century, broadband communication, which is synonymous with the Internet, also changed the IT landscape. Large amounts of information processing are now being outsourced by tangible production industries. Thus, it can be said that ICT has created and expanded the scope of the information processing industry.

Structures of information processing have also changed, and distributed computing, such as grid computing, has become possible [NII]. Cloud computing has also been invented, and production has taken a new shape, with ITeS has become applicable to all production industries.

1.4   Service Creation and ICT

ICT contributes to the promotion of social development. In the following, illustrations of the use of ICT to create the production of services are presented. Some cases are described to demonstrate the services that have become the core source of competitiveness.

1.4.1   ICT and Service Production

Traditional views of information systems usually regard ICT as a process that is used to produce "services within the company." Recent developments have enabled ICT to produce "services" as products for sale. In addition, there are new kinds of "services" in logistics networks, which have created a new distribution industry. Conventional service industries, such as financial services, also welcomed the creation of new services.

1.4.1.1   Relationship Between ICT and Production

Visualization of the entire production process can be regarded as "services within the company." Traditional MIS theory has focused on the corporation's generation and processing of information about production processes, such as inventory, accounting, finance, and the like. These processes are enabled by ICT. Hence, the relationship between ICT and production is simply described as ITeS.

1.4.1.2   The "Clicks and Mortar" Model

At the early stage of Internet use, opening a homepage was considered as the main function of ICT. ITeS at this stage was intended to assist in the creation of websites for one's company or for others. For example, Yahoo! gained success in providing indexes to websites and made portal sites for introducing these websites. Another successful example at this stage was Amazon.com, which opened a web-based store instead of a physical one.

Subsequently, companies that had physical stores decided to open online stores, creating the "clicks and mortar" business model. As online stores and the system for deliveries and payments are established, the range of products sold expanded to include intangible goods, such as software, music tracks, movies and books, thus marking the era of digital content.

In the early twenty-first century, ICT has become a powerful tool in providing various services, with ITeS as an important part of the industry. Web 2.0 was also introduced since the production of services was mainly done through web-based technologies. The evolution of technology lowered the entrance barriers to ITeS, paving the way for the creation of RSS feeds, Facebook, Twitter, and other types of SNSs.

1.4.1.3   Online Gaming

Online games have emerged from the convergence of videogames and SNSs. Social Network Games that allow users to enjoy themselves and interact with one another, is enabled by ICT. Market expansion has been remarkable. In contrast, the market for home videogame devices has contracted.

The difference between the two exists in the way services are provided. Online games depend on the communication network (i.e., the Internet), whereas video games depend on the sales of devices and software. Users simply want enjoyment, and as long as the quality is the same, they are indifferent to the devices involved. The competitive strength of online games comes from its responsiveness to user preferences. For instance, providers of online games survey their products 24 h a day, allowing them to make adjustments is response to players' needs.

1.4.1.4   Distribution Process

ICT has made possible new processes of distribution. Third Party Logistics (3PL) is one such case. With the development of the visualization of production, exchanges of data became possible, first internally, then externally. Consequently, distribution networks now carry a wide range of products from various companies.

In Japan, one of the popular services provided by online stores is the "Daibiki" (pay on delivery) system. Under this system, shoppers buy items at an online store and pay for them upon delivery. Payment collection is done by the delivery person, and is made possible through ICT. Real-time parcel tracking and payment, in turn, are made possible by a specifically developed terminal.

1.4.2   Service as the Source of Competitiveness

The following discusses services as the source of competitiveness. ICT produces services in production and sales industries.

1.4.2.1   Values Created by ICT

Digitization has paved the way for products, such as large LCD TV sets, which are types of commodity whose production used to be a competitive advantage of Japanese corporations. Fine tuning is now no longer required in the production process, because only one IC chip is needed to tune the TV. In the future, the production of these general goods will ultimately become fully automated.

With regards values, the case of high-end TV is a noteworthy example, and in terms of market share, Samsung is widely considered the industry leader. In fact, in September 2012, Hitachi will stop its production line in Japan, and Panasonic has also decided to stop producing high-end large TV sets. The advantages of Samsung are said to be based on the foreign exchange rate. However, this may not be an accurate view. It is product design, together with high quality, which gained consumers' support for its line of products. For example, Hitachi's high-end TVs are equipped with recording functions, but such features are not enough for the brand to gain a large number of consumers. This can be analyzed in relation to the service of recording TV programs, which does not create enough competitiveness. Looking at recording services, one will realize that there is already a large market for dedicated recorders.

1.4.2.2 ICT to Create Value in Distribution

The concept of distribution includes sales. In other words, value is created through the sales process. For example, Rakuten, one of the largest online market places, charges a fee on stores that use its website. The charge includes not only the use of the website, but also the periodical distribution of advertisement through Rakuten's customer lists, search engine optimization, and advice on the design of store websites. Rakuten provides the payment system and distribution support system. Considering the distribution process as a whole, one can see that it is impossible for a small store to operate on its own. As a solution, the web-based service provided by Rakuten is similar to what is provided by a company that manages a large shopping mall.

1.4.2.3 ICT to Create Value in Financing

One of the most important conditions of development is the financing infrastructure. In 14 out of 53 countries, the average national per capita income exceeded US $3,000. Ogimoto et al. (2011) and Kihara et al. (2011) illustrate the financial environment supported by ICT. Among emerging economies, Africa has a large potential to develop. Financing in African countries faces a lack of communication infrastructure. However, the use of mobile phone networks has quickly expanded in recent years; thus, through the use of mobile phone networks and smartphones, "branchless banking" has developed.

Using Short Messaging Services (SMS) and other smartphone functions, various entities can now provide financing services in Africa without having to establish bank branches. The entities involved are banks, mobile phone carriers, non-profit organizations (NPOs), and microfinancing institutions.

Microfinance first became famous in 1995 when Grameen Telecommunication Corporation (GTC) was set up in Bangladesh. To facilitate further success, technological advancements in both broadband mobile phones and smartphones have enabled financing services to penetrate a new geographical area.

1.4.2.4 Service Provision as Production Activities

Providing services is, itself, an act of production, and contributes to an increase in gross national product (GNP). More importantly, the concept of production needs to incorporate the creation of services, because it is the source of competitiveness and the source of value creation. Reponen et al. (2003) focused on the IT-enabled customer relationship management, because it can be the source of competitiveness of a company and the source of additional value. The term ITeS in some cases only means outsourcing. For example, many companies in the United States have outsourced their call center operations to India where there is a large number of English speakers who can be paid at cheaper rates than their counterparts in the United States. In addition, there is a convenient time difference between the two countries, which allows Indian companies to complement the office hours of their US partners. ITeS, however, should not refer solely to outsourcing. Instead, it should be the source of the growth of national wealth.

1.5   Relationship Between Service and ICT in the Twenty-First Century

As mentioned in the previous sections, ITeS in the twenty-first century has become an important source of economic growth. In the following sections, some notable cases from ITeS and future perspectives are presented.

1.5.1   Innovations Brought by ITeS in Rural Areas

Globalization has been accelerated through various factors, one of which is the development of ICT. For example, the penetration of ICT into the daily lives of rural Japanese has been promoted by the development of broadband mobile phone networks. The following provides discussions of ITeS in the rural context.

1.5.1.1   ITeS and Production in Rural Areas

A typical production model of ITeS in rural areas is known as the "Sixth industry model" (Odagiri 2011). The term, which relates to the number six, comes from the synthesis of "primary industry" plus "secondary industry" plus "tertiary industry" $(1 + 2 + 3 = 6$, hence, "Sixth industry").

In rural areas, primary industries such as fishery and farming have remained as major industries. In the past, it used to be difficult for these sectors to reach consumer markets in metropolitan areas. Therefore, they had to rely on organizations such as *Zen-Noh*, which owned processing facilities and a large distribution channel.

By using ICT, each producer has now gained better access to the market in two ways. One is physical access via ICT, and the other is the access to information on the market in metropolitan areas. Therefore, instead of relying on a middle person, the producers are now capable of producing, processing, and selling their own produce. This one stream process demonstrates the Sixth industry model.

In order to share information, SNSs are also widely used in the Sixth industry model. Information sharing through word-of-mouth recommendations has also become a major driving force in the industry. If SNSs enhance relationships between human beings, and if service is defined as something human beings obtain in a relationship, then what SNSs produce can be considered as services within the Sixth industry model. Hence, SNS may well be the driving force of ITeS.

1.5.1.2 A Case of Distribution Innovation

The case of a small fishery base in the city of Ozu in Ehime prefecture, Japan, is a good example of how ITeS contributes to the development of a rural area. Less than 100 households live in this area, and near the port, there is a fish store that sells only natural varieties of fish (as opposed to raised varieties). A store called *Hamaya* sells fresh fish at high prices over the Internet, and has a net store within Rakuten's e-market place. Hamaya used to be the leader in selling Torafugu, a very expensive kind of Japanese pufferfish. Even though fishermen have been catching this kind of fish in this region for a long time, they have always brought the fish to Shimonoseki, where there is a large dedicated market for Torafugu. When ICT became readily available, *Hamaya* opened an online store, thus expanding the market.

The development of ICT infrastructure, however, was not enough for *Hamaya* to be successful; in fact, the store was only able to start shipping the product after the distribution network of Cool Takkyubin reached its place. Takkyubin is a fast package delivery service, which carries various parcels, including golf bags, ski gear, and suitcases sold by Yamato Holdings. Cool Takkyubin is a refrigerated delivery service that is available to anyone almost anywhere in Japan for an additional cost of less than 1,000 yen. The Takkyubin network was built with the help of ICT in order to minimize delivery time. Thus, using this network, delivery is now being done overnight from *Hamaya* to any consumer in Tokyo.

The distribution service, such as Cool Takkyubin, is enabled by ICT. Considering that *Hamaya*'s business (i.e., the electronic sale of fresh fish) is only possible with the development of such distribution services, it can be easily seen that ITeS plays a key factor in rural development.

1.5.2 Technologies and Dreams for the Future

ITeS has a large growth potential, and being able to tap such potential can produce great benefits in the future. ITeS will contribute to the betterment of human welfare. As information devices continue to evolve, dreams will come true. The following presents the prospects of ITeS.

1.5.2.1   New Services Produced by IT

Nowadays, there are many scenarios where services are born with no physical transactions involved. For example, education, medical services, community services, and financial services produce intangible software instead of hardware. In the field of education, the prevalence of Internet usage has transformed the academic environment. Medical services are characterized not only by tangible operations, such as long-distance or remote surgery using robotic arms, but also intangible operations such as shared use of patients' treatment histories. In community services, regional SNSs are focused on the enhancement of human relationships in a certain geographical area. Financial systems, such as local currency systems can be built as a part of this IT infrastructure. Moreover, their collaboration with incumbent banking systems can also be enhanced through ITeS.

These are the areas of new services that will emerge and evolve. The upgrade of devices will also continue, making it possible for ITeS to expand the scale and scope of various services.

1.5.2.2   "Intelligence" Crosses Borders

ITeS crosses boundaries. Hidehiko Sanada, Professor Emeritus of Osaka University, once explained the nature of corporate activities by focusing on four flows in the business. They are "Knowledge flow," "Commercial activity flow," "Physical Distribution flow," and "Financial flow." According to him, "those flows summarize the process of business. It is the process of knowing what a corporation can do (seeds), understanding how to respond to the market wants (needs), physically producing and delivering and gaining the money." The application of IT began from "Commercial activity flow," such as CIM and POS, and then followed by "Physical distribution," which enabled Takkyubin services. Electronic payment transactions are undergoing. The rest is "Knowledge flow," which involves activities such as R&D and marketing (Sanada 2001). He delivered his lecture at about the same period when Reponen et al. were conducting research on Global Customer Service.

About a decade later, marketing with accumulated consumer logs has become one of the most popular services in cyberspace. Tele-collaboration between researchers in R&D is another. Integrating different datasets and profiles of SNS users to come up with a more detailed customer database has also become widespread business practice. Such activities are often conducted across geographical borders.

To reflect the current environment, "Knowledge flow" may also be referred to as "Intelligence flow." The Oxford Dictionary defines "knowledge" as "facts, information, and skills acquired through experience or education; the theoretical or practical understanding of a subject," whereas "intelligence" is defined as "the ability to acquire and apply knowledge and skill." Evidently, designing new ITeS that can cross borders requires creativity in applying relevant knowledge obtained.

1.6 Conclusion

This chapter discussed the concepts and theories related to ITeS. The ITeS concept reflects an important stage in the development of information systems. Descriptions of Nolan's stage theories and the work of Shimada and Takahara have been presented. Studies conducted by Reponen et al. have also been revisited. The concept of ITeS used to focus on customer relationship management. The review of Reponen et al. demonstrated how the studies of eCRM have been concluded.

Furthermore, this chapter reviewed the relationship between IT and ICT. The conversion from analog to digital technology has also been described, emphasizing the importance and social implications of digitization.

This chapter also looked into the technological factors influencing ITeS. Service is embedded in the product and is delivered through the production process. Visualization can be done through the evolution of digitization. In addition, the competitiveness of products is rooted in how service factors are incorporated in them.

Finally, ITeS in the twenty-first century has been discussed. Predictions of future development in such areas as education, medicine, community and financial services are also presented. To conclude, this chapter presented the concept of business flows and its development in the cross-border provision of services.

Acknowledgments This is the produce of Matsuyama University's Regional Research Center Project (Chiiki Kenkyu Center Project, 2008–2011) and part of this research has been funded by MEXT (Ministry of Education, Culture, Sports, Science and Technology, Japan) Programme for Strategic Research Bases at Private Universities (2012–2016) project "Organisational Information Ethics" S1291006, and Grants in Aid (KAKEN) B12022425.

References

Dan Y, Okada H, Uesugi S (2009) Application of production possibility frontier model to IT-enabled services. In: Proceedings of the ITS Asia-Africa-Austrasia regional conference 2009, Curtin Business School, Perth

Gibson CF, Nolan RN (1974) Managing the four stages of EDP growth. Harvard Bus Rev 52 (1):76–88

http://oxforddictionaries.com/definition/intelligence

http://oxforddictionaries.com/definition/knowledge

Karimi J, Somers TS, Gupta YP (2001) Impact of information technology management practices on customer service. J Manage Inf Syst 17(4):125–158

Kihara Y et al (2011) Afurica no Kinyu Sekuta Chu (African financial sector 2, in Japanese). NRI, Tokyo

McFarlan FW, Nolan RL, Norton DP (1973) Information systems administration. Holt/Rinehart/ Winston, New York

Miyakawa T (1994) Keiei Joho Sisutemu (Management information system, in Japanese). Chuokeizaisha, Tokyo

Nolan RL (1973) Managing the computer resource: a stage hypothesis. Commun ACM 16 (7):399–405

Nolan RL (1979) Managing the crisis in data processing. Harvard Bus Rev 57:115–126

Nolan RL (1982) Managing the data resource function. West Publishing, St. Paul

Nolan RL, Croson DC (1995) Creative destruction: a six-stage process for transforming the organization. HBR Press, Boston

Odagiri T (2011) Rural regeneration in Japan. Centre for Rural Economy Research Report. University of Newcastle upon Tyne

Ogimoto Y et al (2011) Afurica no Kinyu Sekuta Jou (African financial sector 1, in Japanese). NRI, Tokyo

Okada H, Uesugi S (ed) (2009) J Inform Reg Stud 1(1). Matsuyama University, Matsuyama

Okada H, Uesugi S (ed) (2010) J Inform Reg Stud 2(1). Matsuyama University, Matsuyama

Okada H, Uesugi S (ed) (2011) J Inform Reg Stud 3(1). Matsuyama University, Matsuyama

Okada H, Uesugi S (ed) (2012) J Inform Reg Stud 4(1). Matsuyama University, Matsuyama

Pugh EW (1995) Building IBM: shaping an industry and its technology. The MIT Press, Cambridge

Reponen T (ed) (2003a) Information technology-enabled global customer service. Idea Group, Hershey

Reponen T (2003b) Service perspective of business information systems. In: Reponen T (ed) Information technology-enabled global customer service. Idea Group, Hershey, pp 1–8

Sanada H (2001) Evolution of network and e-society. In: 4th OFC lecture. http://www2.econ. osaka-u.ac.jp/ofc/ofclecture4.htm. Accessed 15 Mar 2012

Shimada T, Takahara Y (1993) Keiei Joho Sisutemu (Management information system, in Japanese). Nikkagiren, Tokyo

Treebhoohu N (2012) Promoting IT enabled services. Commonwealth Secretariat, London

Tsokota T (2011) The feasibility of setting up information technology-enabled services. LAP Lambert Academic Publishing, Saarbrücken

Uesugi S (2008) Bridging between real and virtual – technologies to advance ITeS. In: IEEE/IPSJ international symposium on applications and the internet. doi ieeecomputersociety.org/ 10.1109/SAINT.2008.48

Ulmann B (2010) Analogrechner. Oldenbourg Wissensch.Vlg

Chapter 2
The Effects of Similarities to Previous Buyers on Trust and Intention to Buy from E-Commerce Stores: An Experimental Study Based on the SVS Model

Tetsuro Kobayashi and Hitoshi Okada

2.1 Introduction

The spread of business-to-customer e-commerce in recent years has led to a growing body of studies on the role of trust in Internet shopping (Fogg and Tseng 1999; Fogg et al. 2001; McKnight 2001; McKnight and Chervany 2002; McKnight et al. 2002; Gefen et al. 2003a, b; Salam et al. 2005). Trust concerns people's perception of a website's trustworthiness in the face of social uncertainties that remain despite institutional and technological protective structures. Although institutional and technological advancements continue to secure structural assurance (McKnight and Chervany 2002; McKnight et al. 2002; Gefen et al. 2003b), system trust (Grabner-Kräuter and Kaluscha 2003), and calculative-based beliefs (Gefen et al. 2003b), lack of trust is still a strong inhibiting factor to the spread of e-commerce (Wang et al. 1998; Hoffman et al. 1999; Jarvenpaa et al. 1999; Gefen and Straub 2004). Therefore, along with the sophistication of the definition of trust in e-commerce (McKnight and Chervany 2002; Mayer et al. 1995; McKnight et al. 1998), many studies have investigated the question of how to build customer trust under the uncertain conditions of online financial transactions (Grabner-Kräuter and Kaluscha 2003; Corbitt et al. 2003; Koufaris and Hampton-Sosa 2004; Lim et al. 2006).

There are several research approaches to investigating the trust-building methodologies in an e-commerce environment. First, several studies have focused on the sophistication of contents and interfaces of e-commerce websites. For example, Fogg and colleagues have examined the effectiveness of information in aiding judgments of a website's trustworthiness when potential customers browse websites (Fogg and Tseng 1999; Fogg et al. 2001, 2003; Fogg 2002). Similarly, other studies have applied the technology acceptance model (TAM) (Davis 1986,

T. Kobayashi (✉) • H. Okada
National Institute of Informatics, 2-1-2 Hitotsubashi, Chiyoda-ku, Tokyo, Japan
e-mail: k-tetsu@nii.ac.jp; okada@nii.ac.jp

S. Uesugi (ed.), *IT Enabled Services*,
DOI 10.1007/978-3-7091-1425-4_2, © Springer-Verlag Wien 2013

1989), which focuses on the perceived ease of use and usefulness of websites (Gefen et al. 2003a, b; Pavlou 2003). These approaches have investigated how to transcend the signals of trustworthiness of the websites by increasing the sophistication of the interface and enriching the contents of e-commerce websites.

On the other hand, in addition to information on the website, there are studies that evaluate the effectiveness of information and reputation that exist *outside* the focal e-commerce website in building trust among potential buyers (Lim et al. 2006; Lowry et al. 2008). Although it is obvious that both approaches are necessary to understand how to build trust among potential buyers, empirical research on the latter is scarce in comparison with that on the former. Therefore, this study sheds light on the effectiveness of external information from a different perspective from those of previous studies. External information is defined as the information that is not present on the focal e-commerce website but can be useful in judging the trustworthiness of the store.

In this study, we first discuss the limitations of trust-building approaches based on traditional social psychology that focuses only on the internal components of websites (henceforth referred to as "traditional social psychological approaches"). Then we present empirical evidence that external information on salient value similarities (SVSs) with previous buyers is a critical factor in building trust in e-commerce stores among potential buyers. Through experimental manipulation of SVSs, we draw valid and rigorous inferences about causal relationships that are not available from correlational studies based on questionnaire surveys.

2.2 Theoretical Development

2.2.1 Limitations of Traditional Social Psychological Approaches

Previous studies have demonstrated the diverse and complex nature of trust in the context of e-commerce (McKnight and Chervany 2002; McKnight et al. 2002; Doney and Cannon 1997; Corritore et al. 2003). In fact, there is no clear consensus on the definition of trust so far (Kee and Knox 1970; Driscoll 1978; Cook and Wall 1980; Scott 1980). As McKnight et al. (1998) note, the word "trust" is so confusing (Shapiro 1987) and broad (Williamson 1993) that it almost defies careful definition (Gambetta 1988). In particular, because e-commerce studies are interdisciplinary, there are different sentences in the "grammar" of trust (McKnight and Chervany 2002), which leads to the difficulty in achieving consensus. Despite this difficulty, however, recent studies have tried to build consensus on the definition of Mayer et al. (1995). Mayer et al. (1995) defined trust as the willingness of a party to be vulnerable to the actions of another party based on the expectation that the other will perform a particular action important to the trustor, irrespective of the ability to monitor or control the other party. Based on this definition, Mayer et al. (1995) conceptualized perceived competence (i.e. ability), benevolence, and integrity as

Fig. 2.1 Traditional social psychological model

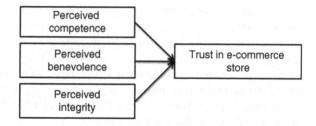

the antecedents of trust (Fig. 2.1). Competence refers to skills, abilities, and characteristics that enable a party to have influence within a specific domain. Benevolence is the extent to which a trustee is believed to want to do good to the trustor, aside from an egocentric profit motive. Integrity is defined as the trustor's perception that the trustee adheres to a set of principles that the trustor finds acceptable.

According to the definition of trust by Mayer et al. (1995), potential buyers evaluate information on the competence, benevolence, and integrity of e-commerce websites respectively and finally judge whether they trust the website. In fact, many previous studies on trust in e-commerce have conceptualized models consistent with Mayer et al. (1995).

For example, using large-scale social surveys, Fogg et al. (2001) identified 300 internal components that make websites credible (see also Fogg (Davis 1986)). These components were reduced to 51 items classified under seven different subscales: real-world feel, ease of use, expertise, trustworthiness, tailoring, commercial implications, and amateurism. It should be noted that all these concern perceptions of site components.[1] Likewise, studies by Gefen et al. (2003a, 2003b); Salam et al. (2005), which employ the TAM (Davis 1986, 1989) to estimate trust based on a website's perceived ease of use and usefulness, also concern the internal components of a website. Consistent with the model of Mayer et al. (1995), these studies all indicate that it is important for e-commerce stores to communicate their trustworthiness accurately through the website's design, ease of use, and expertise, to gain customers' trust in e-commerce.

Originally, the approach that focuses on perceived ability, benevolence, and integrity was derived from findings on persuasive communication in social psychology. Hovland and Weiss' (1951) classic experiment showed that there is a high level of trust in a message if the credibility of its source is high. The components of credibility are (1) a perception that the source has expert knowledge, experience or qualifications, and (2) the perception that the source is an honest person with a benevolent intent to convey a message fairly (Hovland et al. 1953; McGinnies and Ward 1980). That is, if the receiver easily perceives competence, benevolence, and integrity, he or she will trust the sender.

---

[1] To be exact, the trustworthiness scale indirectly considers the effects of external information, such as the presence or absence of links to external sites. However, because these links are presented on the website, it is appropriate to count them as internal components.

However, there is an important presupposition to the traditional social psychological model. That is, it is assumed that the receiver of a message can obtain sufficient and accurate information to evaluate the competence, benevolence, and integrity of the sender. If the receiver was unable to obtain this information, or the information was not credible even if obtained, s/he would be unable to judge whether the sender may be trusted. The question is whether the internal components of a website, such as those presented by Fogg et al. (2001), are sufficient to allow a buyer to evaluate a seller's trustworthiness.

Compared with brick-and-mortar stores, transactions at online stores generally involve greater social uncertainties because buyers are unable to confirm directly the quality of a product or engage in face-to-face communication with the seller (Reichheld and Schefter 2000). Therefore, it is difficult to gather the sufficient and credible information regarding the seller's competence, benevolence, and integrity that is normally accessible at brick-and-mortar stores where customers can talk with retailers and actually see the products (Kollock 1999; Gefen 2000). Traditional social psychological approaches can be regarded as attempts to overcome this lack of information through enriched, sophisticated website content and design. However, these improvements may not provide the consumer with sufficient signals concerning competence and benevolence/integrity. The reason is that, at least compared with brick-and-mortar stores, the components of e-commerce stores can be much more easily imitated and reproduced, as has been evident in rampant online phishing. Even if trust-building components are identified, these alone cannot be relied upon as stable signals of a website's trustworthiness in the long term because they are easily imitated and reproduced. This indicates that we need to be careful in applying the definition of trust by Mayer et al. (1995) because, as these authors noted, "this model is focused on trust in an organizational relationship, and its propositions may not generalize to relationships in other contexts." Mayer et al. (1995) focused on the relationships in organizations such as between employers and employees and between supervisors and their subordinates. Their relationships are normally based on face-to-face interactions where they can obtain ample clues about the competence, benevolence, and integrity of others, which is not necessarily the case in e-commerce situations.

In summary, in e-commerce transactions where uncertainty caused by information asymmetry is high, trust-building strategies based on traditional social psychological approaches may malfunction because internal components inside the website that can be easily forged or imitated at lower cost cannot fully transmit effective and costly signals. In this situation, enrichment and sophistication of internal components of the website do not necessarily guarantee greater consumer trust. In other words, having the appropriate internal components is a necessary but not sufficient condition for gaining a customer's trust. This suggests that we need to look beyond a website's internal components to understand customer trust building in e-commerce.

2.2.2   Effectiveness of External Information

Considering the limitations of traditional social psychological approaches, we next examine how external information may be used by potential buyers to evaluate a seller's competence, benevolence, and integrity. For example, employing a large-scale social survey, Fogg et al. (2003) refined component approaches and identified "name recognition and reputation" as one of 18 categories of information clues for respondents. Trust in a website increases if the site operator's name is well known in the real world, and this effect is especially apparent in e-commerce stores. In fact, Fogg et al. (2003) recognized that it is impossible to control reputation and other such external information solely through enrichment and sophistication of the internal components of a website and called for more research on the effect of external information on trust.

Noting the importance of external information, some previous studies have investigated the effect of third party certificates, consumer feedback, and advertising reputation. However, these studies have mainly focused on the trust-building effects of such information when it is presented on the website. For example, Cheskin Research (1999, 2000) evaluated the effectiveness of TRUSTe seals in trust building (see also McKnight and Chervany (2001)). Similar approaches have been applied to BBB online (Cheskin Research and Studio Archetype/Sapient 1999; Cheskin Research 2000; McKnight and Chervany 2001), WebTrust (McKnight and Chervany 2001; Kover et al. 2000a, b), and VeriSign (Cheskin Research and Studio Archetype/Sapient 1999; Cheskin Research 2000; McKnight and Chervany 2001). Other studies have investigated the effect of advertising reputation (Cheskin Research and Studio Archetype/Sapient 1999; Cheskin Research 2000; McKnight and Chervany 2001; Jarvenpaa et al. 2000), customer feedback (Lim et al. 2006, 2001), and portal affiliation (Lim et al. 2006). Although these previous studies concern the effectiveness of the credibility of third parties or customer feedback rather than self-report by the e-commerce store, they are still similar to the traditional social psychological model in that they focus on third party certification or customer feedback presented on the e-commerce website.

Although these studies are important in understanding how website design and information presentation affect trust building, it must be noted that potential buyers do not judge the trustworthiness of an e-commerce site only from the information presented on it. When people make purchases via the Internet, they can collect external information on e-commerce stores through any number of search engines, word of mouth sites, bulletin boards, and blogs. To build trust in e-commerce, it should be important to employ these wide networks of reputation and recommendation information appropriately. As mentioned above, it is difficult for potential buyers to penetrate the website's disguise with fake certification seals or forged customer feedback because of the information asymmetry in the e-commerce situation. When the internal components of the website cannot serve as stable signals of trustworthiness, potential buyers may also search for information outside the website to judge its trustworthiness. That is, potential buyers not only gather clues about competence, benevolence, and integrity of the website in a bottom-up way, but it is also usual for them to judge whether to trust a website based on the information outside it.

However, there are a surprisingly small number of studies that investigate the effect of external information on potential buyers' trust. This leads to a need for a clear process model to explain what and how external information about e-commerce stores influences trust among potential buyers. Rather than the traditional social psychological model, which assumes that information on a website's competence and benevolence/integrity are readily available, we must adopt a bounded rationality model that accounts for the constraints of social uncertainties characteristic of e-commerce stores. In this study, we employ the SVS model to investigate the trust-building process in e-commerce stores.

2.2.3 The SVS Model of Trust and Its Application to E-Commerce

The SVS model of trust was originally developed from risk perception studies. According to Earle and Cvetkovich (1995), people trust others when they perceive that they have the same salient values. For example, citizens trust a particular risk management organization if they feel that the organization's priorities in approaching and solving a problem (salient values) are similar to their own. Based on shared salient values, they entrust the organization with decision-making power.

In contrast to the traditional social psychological model, the key feature of the SVS model is that it explains trust in situations with insufficient clues for a person to evaluate the competence or benevolence/integrity of another person or an organization. When people cannot directly confirm the competence or benevolence/integrity of others, they focus on similarity in salient values. If they conclude that similarity exists, they will "entrust" these others with decision making.

The SVS model is entirely different from the traditional social psychological model in that, rather than competence and benevolence/integrity being treated as antecedents of trust, perception of SVS increases trust, leading to increased perception of competence and benevolence/integrity. Many empirical risk management studies have supported the SVS model (Earle and Cvetkovich 1997; Siegrist and Cvetkovich 2000; Siegrist et al. 2001, 2003, 2005; Earle 2004; Poortinga and Pidgeon 2006). Furthermore, direct comparisons of the SVS and traditional social psychological models in the area of risk management have indicated that the SVS model fits the data better than the traditional social psychological model (Cvetkovich and Nakayachi 2007).

The SVS model of trust can be applied to trust building in e-commerce, where information on competence and benevolence/integrity is not easily transmitted. As discussed above, it can be difficult for users to gather information about the competence and benevolence/integrity of an e-commerce website when they decide whether to trust it. Given these social uncertainties surrounding the trustworthiness

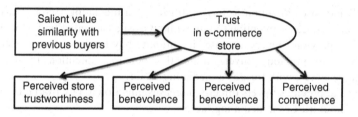

Fig. 2.2 Salient value similarity model

of a site, external information is important. According to the SVS model of trust, people base their decisions on the judgments of others with whom they share salient values. If we define information about the salient values of previous buyers as external information, it follows that if potential buyers perceive that they share salient values with previous buyers, they will trust them and consequently their judgment of the trustworthiness of a particular e-commerce store. By definition, previous buyers have already decided to trust an e-commerce store and have made a purchase. If a person's salient values are very similar to those of a previous buyer, his or her trust in, and intention to buy from, the e-commerce store also increases. The SVS-based trust-building model we use in this study is illustrated in Fig. 2.2. Based on this model, trust in this study is constitutively defined as a psychological construct comprised of perceived trustworthiness, benevolence, integrity, and competence.

Based on the theoretical development discussed above, we have two hypotheses to test in this study.

H1: SVS to previous buyers increases trust in e-commerce stores.

Trust in e-commerce is a complex concept that cannot be measured with any single indicator (McKnight 2001; McKnight and Chervany 2002; McKnight et al. 2002). In this study, we take store trustworthiness, benevolence, integrity, and competence as four indicators of trust-related variables.

In addition to manipulating SVS in our experiment, we also examine the effect of basic attribute similarities that potential buyers may have with previous buyers. Zucker (1986) argues that similarities in basic attributes, such as birthplace and race, can produce mutual trust. It is possible that this type of trust-building process is also at work in an e-commerce context. In this study, by attributes we mean sex and age group. If similarity leads to greater trust in e-commerce stores, is it limited to similarity of salient values, or does it include wider similarity of social attributes? We must answer this question to clarify the scope of our proposed model.

There are also significant implications for marketing if SVS to previous buyers is found to have a substantial effect on intention to buy. Although the SVS model of trust predicts that SVS increases trust, it does not predict subsequent behavior. Even if SVS to previous buyers increases intention to buy, with our present knowledge we have no way of predicting whether trust acts as a mediating factor or if SVS directly affects intention to buy. Therefore, we divide our second hypothesis, that SVS to previous buyers increases intention to buy, into two competing corollaries: SVS *indirectly* increases intention to buy with trust as a mediating factor (H2-1), and SVS *directly* increases intention to buy without trust as a mediating factor (H2-2).

H2: SVS to previous buyers increases intention to buy from e-commerce stores.

H2-1: SVS to previous buyers indirectly increases intention to buy from e-commerce stores, with trust as a mediating factor.

H2-2: SVS to previous buyers directly increases intention to buy from e-commerce stores, without trust as a mediating factor.

2.3   Methodology

2.3.1   *Presurvey and Selection of Subjects*

The subjects of this experimental survey were Japanese adults who were recruited through "goo Research," a web-based survey service of NTT Resonant Inc (Tokyo, Japan).[2] We first conducted a presurvey over two periods: January 29th–31st and February 4th–8th 2009. Our presurvey of 75,000 goo Research registrants produced 2,151 valid responses, limited to adults aged from 20 to 39 years old. This was to ensure a sufficient spread of ages for attribute dissimilarity, to be discussed below.

The primary information we gathered in the presurvey consisted of (1) past e-commerce experience, (2) consumer values related to food product e-commerce and their subjective importance (salient values), and (3) basic social attributes. We first selected respondents who had made a past e-commerce purchase of at least one of (1) crab or other seafood, (2) meat, (3) side dishes, (4) rice, or (5) vegetables or fruit. The purpose of limiting the survey to respondents with food purchase experience was to increase the sense of reality in the following experimental survey.

We next used a four-point bipolar scale to measure ten different consumer values related to food product e-commerce. We also used a four point scale ("important," "somewhat important," "somewhat unimportant," and "unimportant") to measure the subjective importance attached to these consumer values by respondents. This information was used as point of reference in the experimental survey when presenting SVSs. Specifically, we adopted the two items shown in Table 2.1 because they had comparatively symmetric distributions and high average values for subjective importance. We also selected respondents who ranked these two items as "important" or "somewhat important" to ensure the success of the manipulation in the experimental survey.

After selecting subjects, we conducted the experimental survey using a two-factor between-subjects design with random assignment. Salient consumer value and attributes were each tested at three levels: similar attributes/values, dissimilar attributes/values, or no attributes/values. For each of the nine conditions

---

[2] This study was supported by a Grant-in-Aid for Scientific Research (PI: Hitoshi Okada, # 20402034).

Table 2.1 Salient consumer values used for salient value similarity manipulation

		Close to A	Somewhat close to A	Somewhat close to B	Close to B	
Salient consumer value 1	A: Select inexpensive and economical products	10.12	39.41	38.88	11.58	B: Select high-quality products even though they are expensive
Salient consumer value 2	A: Select best products by spending considerable time to ensure full satisfaction	24.1	26.9	44.74	4.26	B: Select satisfactory products without spending much time on it

Percentage (%) of experimental survey subjects (N = 751)

$(3 \times 3 = 9)$, we solicited 180 respondents for a total of 1,620 survey subjects. The distribution of salient consumer values was equal in each of the nine conditions. That is, 25 % of respondents in each condition fell into each of the four combinations of salient consumer values 1 and 2, with possible responses A or B. The combinations were: (1) A, A; (2) A, B; (3) B, A; and (4) B, B.

2.3.2   Experimental Survey and Manipulation Check

We conducted our experimental survey on February 10th and 11th 2009. The procedures were conducted in four steps.

Step 1: Lead-in to the Scenario. The following text was presented to all subjects.
"You have a craving for crab after seeing on television that it is in season. However, you've been very busy recently and don't have time to go to the store. You decide to buy some crab from an online store instead. After browsing around, you narrow down your choices to a number of stores selling snow crab that you think you would be satisfied with in terms of price and taste. One of the stores you are considering is shown on the next page. Please carefully look over the information on this site before moving on to the next page."

Step 2: Mock Site. The mock site of an e-commerce store that sells crab was presented to all subjects.[3] The e-commerce site included pictures, price, brief description about the quality of crabs, weight, pull date, and place of origin, all of which are typical internal components of e-commerce sites selling crabs in Japan.

Step 3: Presentation of Salient Consumer Values and Attributes of Previous Buyers. The following text was presented to subjects based on salient consumer values measured in the presurvey.

---

[3] The mock e-commerce site shown to subjects is available from the authors upon request.

"According to "goo Research," a survey of previous buyers indicates that this shopping site is used mostly by [attribute 1 (sex)] in their [attribute 2 (age group)] who [salient consumer value 1]. Recently it has also become popular among people who [salient consumer value 2]."

Subjects assigned to the similar attribute condition were informed that many previous buyers were of the same age group and sex as themselves. Subjects assigned to the dissimilar attribute condition were informed that many previous buyers were over 30 years of age and of the opposite sex. Those assigned to the similar salient consumer value condition were informed that many previous buyers shared two of the respondent's salient values measured in the presurvey. In the dissimilar salient consumer value condition, subjects were informed that many previous buyers endorsed values opposed to those that the subjects had indicated in the presurvey. For example, to women in their 20s who answered A for salient consumer value 1 and A for value 2 in the presurvey and were assigned to the similar attributes and salient consumer values condition, we presented the following information about previous buyers.

"According to "goo Research," a survey of previous buyers indicates that this store is used mostly by women in their 20s who select inexpensive and economical products. Recently it has also become popular among people who select the best products by spending considerable time to ensure full satisfaction." (Similar attributes and similar salient consumer values condition)

To men in their 30s who answered B for salient consumer value 1 and A for value 2 in the presurvey assigned to the dissimilar attributes and dissimilar salient consumer values condition, we presented the following information about previous buyers.

"According to "goo Research," a survey of previous buyers indicates that this store site is used mostly by women in their 60s who select inexpensive and economical products. Recently it has also become popular among people who select satisfactory products without spending much time." (Dissimilar attributes and dissimilar salient consumer values condition)

Subjects assigned to the no-attributes condition were only given information on the salient consumer values of previous buyers. Those in the no salient consumer values condition were only provided with information on the attributes of previous buyers. Subjects with neither attributes nor salient consumer values were not provided with any information about previous buyers. Finally, for purposes of checking the manipulation, we calculated the total amount of time spent by each subject on steps 2 and 3.

Step 4: Measurement of Dependent Variables and Manipulation Check. After reviewing the survey results of previous buyers, each of the dependent variables below were measured. Please refer to the Appendix for the scale items.

1. Perceived store trustworthiness
   We used six of the seven items from the store trustworthiness scale (Jarvenpaa et al. 1999) ($\alpha = 0.73$).
2. Perceived benevolence, integrity, and competence

Table 2.2 Factor analysis of all indicators

	Variable	Factor 1	Factor 2	Uniqueness
Trust-related scales	Store trustworthiness	0.86	0.02	0.24
	Benevolence	0.89	−0.01	0.22
	Integrity	0.90	0.02	0.17
	Competence	0.82	0.05	0.28
Transaction intentions	Item 1	0.18	0.75	0.26
	Item 2	−0.05	0.92	0.20
	Item 3	0.14	0.81	0.21
	Item 4	−0.10	0.86	0.34
	Variance	3.99	3.78	
N = 751	Proportion	0.50	0.47	

Method: principal component factor analysis
Rotation: oblique promax

We used slightly reworded versions of three subscales from the trusting belief scale (McKnight et al. 2002), benevolence scale (three items, $\alpha = 0.72$), integrity scale (four items, $\alpha = 0.89$), and competence scale (four items, $\alpha = 0.85$).
3. Transaction intentions
We used a total of four items: one of three items on the transaction intentions scale (Item 1 in the Appendix) (Pavlou and Gefen 2004) and three of four items on the intention to buy scale (Items 2 to 4 in the Appendix) (Stewart 2003) ($\alpha = 0.86$).

A factor analysis (principal component method) of all of the indicators above clearly indicated a two-factor structure (Table 2.2). This result validates our models with two latent variables; i.e. trust in an e-commerce store and intention to buy.

There were 941 responses to the 1,620 requests to participate in this study. We excluded data with low trustworthiness by eliminating those from the fastest and slowest 5 % of respondents in terms of total response time as well as those from respondents who spent less than 10 s or more than 10 min on steps 2 and 3. In addition, we eliminated cases that indicated incomplete manipulation.[4] This left 751 subjects for analysis. An ex posteriori sample size calculation (Westland 2010) indicated that the sample size of 751 subjects is adequate for hypothesis testing using structural equation models.

---

[4] A manipulation check was conducted after the dependent variables were measured. Respondents were asked, "Of the previous buyers who previously bought crab from this store, do you believe that there are many people who are similar to you, or do you believe that there are more people who are dissimilar to you, based on the criteria below?" We eliminated the subjects who responded "dissimilar" to at least one of the two salient values although they were assigned to the similar salient consumer value condition, as well as subjects who responded "similar" to at least one of the two salient values although they were assigned to the dissimilar salient consumer value condition. Sixty-six subjects were eliminated by this manipulation check. This may be because of an interval of up to 2 weeks between the presurvey and experimental survey and possible differences in evaluation criteria used for food products in general in the presurvey and for crab only in the experimental survey.

Table 2.3 Correlation matrix for dependent variable scales

	Store trustworthiness	Benevolence	Integrity	Competence
Benevolence	0.72			
Integrity	0.77	0.77		
Competence	0.65	0.71	0.74	
Transaction intention	0.48	0.47	0.50	0.50

All coefficients are significant at 1 % level (N = 751)

2.4 Results

Table 2.3 shows the correlation coefficient matrix for the dependent variables.

To verify H1 and H2 simultaneously for statistical efficiency, structural equation modeling was employed to determine whether SVS to previous buyers increased trust in the e-commerce store (H1), and whether SVS to previous buyers increased intention to buy from the e-commerce store as an indirect effect mediated by trust, or as an unmediated direct effect (H2). We configured a latent variable to trust in e-commerce store because, as Table 2.3 shows, there were strong correlations among indicators of trust in e-commerce site. We estimated an indirect model with a trust-mediated indirect effect of the exogenous variables (salient consumer value similarity and attribute similarity) on intention to buy, which verifies H2-1, as well as the direct model allowing direct paths from the exogenous variables to intention to buy, which verifies H2-2. By comparing models using goodness of fit indices (GFIs), we investigated the paths through which similarity to previous buyers affects intention to buy.

Figure 2.3 shows the indirect model adopted after the addition of three error covariances. The reference categories of salient consumer value similarity and attribute similarity are the dissimilar salient consumer values and dissimilar attributes conditions. These reference categories are naturally omitted from the model to test the difference in means between each condition with reference categories. Because salient consumer value similarity and attribute similarity were orthogonal, we set all the covariance parameters between the two factors at zero. Table 2.4 shows the indirect model's goodness of fit and standardized coefficients.

As Table 2.4 shows, the GFI, adjusted goodness of fit index (AGFI), and root mean square error of approximation (RMSEA) all indicated acceptable overall goodness of fit. The results of chi-squared tests indicate that there was no statistically significant discrepancy between the data and the model. Furthermore, all the coefficients from trust in e-commerce store to its three indicators are highly significant, which indicates the success of the measurement model of trust (the coefficient of store trustworthiness was set to 1 to fix the scale of a latent variable; i.e. trust in e-commerce store).

Trust was significantly higher in the similar salient consumer values condition compared with the dissimilar condition, which clearly supports H1. On the other hand, similarity in attributes (i.e. sex and age groups) did not show any significant

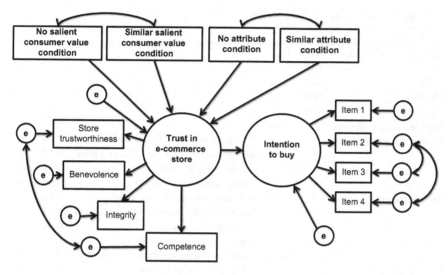

Fig. 2.3 Indirect model

Table 2.4 Estimations for the indirect model

Goodness of fit				
Chi-squared (df = 57)	43.75	n.s		
GFI	0.99			
AGFI	0.98			
RMSEA	0.00			
AIC	103.75			
Standardized coefficients				Beta
Trust in e-commerce store	←	No salient consumer value condition	0.09	+
	←	Similar salient consumer value condition	0.14	**
	←	No attributes condition	0.01	n.s.
	←	Similar attributes condition	0.04	n.s.
Intention to buy	←	Trust in e-commerce store	0.64	**
Store trustworthiness	←	Trust in e-commerce store	0.84	–
Benevolence	←		0.83	**
Integrity	←		0.89	**
Competence	←		0.81	**
Item 1	←	Intention to buy	0.89	–
Item 2	←		0.71	**
Item 3	←		0.89	**
Item 4	←		0.60	**

+ p < 0.10, ** p < 0.01
All the covariances are significant at 1 % level
See Appendix for the scale items of intention to buy

Table 2.5 Estimations for the direct model

Goodness of fit				
Chi-squared (df = 53)	36.65	n.s.		
GFI	0.99			
AGFI	0.99			
RMSEA	0.00			
AIC	104.65			
Standardized coefficients	Beta			
Trust in e-commerce store	←	No salient consumer value condition	0.09	+
	←	Similar salient consumer value condition	0.13	**
	←	No attributes condition	0.01	n.s.
	←	Similar attributes condition	0.04	n.s.
Intention to buy	←	No salient consumer value condition	0.01	
	←	Similar salient consumer value condition	0.07	+
	←	No attributes condition	0.06	+
	←	Similar attributes condition	0.05	
	←	Trust in e-commerce store	0.64	**
Store trustworthiness	←	Trust in e-commerce store	0.84	–
Benevolence	←		0.83	**
Integrity	←		0.89	**
Competence	←		0.81	**
Item 1	←	Intention to buy	0.89	–
Item 2	←		0.71	**
Item 3	←		0.89	**
Item 4	←		0.60	**

+ $p < 0.10$, **$p < 0.01$
All the covariances are significant at 1 % level
See Appendix for the scale items of intention to buy

effects on trust in e-commerce store. Trust in e-commerce store, in turn, had a statistically significant positive effect on intention to buy.

Table 2.5 shows the goodness of fit and standardized coefficients of the direct model (Fig. 2.4) in which we draw direct paths from salient consumer value similarity and attribute similarity to intention to buy. GFI, AGFI, and RMSEA all indicated acceptable overall goodness of fit. The results of chi-squared tests indicate no statistically significant discrepancy between the data and the model. Trust was significantly higher in the similar salient consumer values condition compared with the dissimilar condition as well as in the indirect model. The consistency of the positive effect of salient consumer values similarity across two models indicates robustness of the validity of H1. Attribute similarities did not increase trust in e-commerce store as well as in the indirect model. Furthermore, the direct paths from value similarity and attribute similarity did not show clear positive effects on intention to buy from the e-commerce site. Although two of the four coefficients are marginally significant, these effects are rather weaker than the effects on trust in e-commerce store.

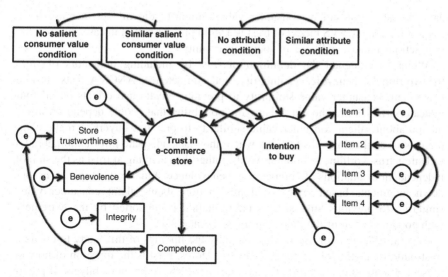

Fig. 2.4 Direct model

Furthermore, there was no statistically significant improvement in overall goodness of fit from the indirect to the direct model (Δchi-squared $= 7.1$, df $= 4$, n.s.), and Akaike's information criterion (AIC) is slightly smaller in the indirect model. This strongly suggests that we should select the indirect model from the viewpoint of parsimony. These results support H2-1, instead of H2-2, that salient consumer value similarity has an indirect effect, via trust in e-commerce store, on intention to buy.

2.5   Discussion and Conclusions

In this study, we indicated the limitations of traditional social psychological approaches in understanding trust building in e-commerce stores and investigated the effects of external information about salient consumer values and attributes of previous buyers on trust in potential buyers. We also conducted an exploratory investigation of the process by which SVS increases intention to buy from an e-commerce store.

As we predicted in H1, salient consumer value similarity to previous users had a causal effect on increasing trust in e-commerce stores. By manipulating salient consumer value similarity, we were able to avoid the ambiguous interpretation common to correlational survey studies. Specifically, it is not possible to interpret our findings such that there was a reverse causal relationship in which people trust e-commerce stores and therefore perceive shared similarities with previous buyers under a false consensus effect (Ross et al. 1977). Sharing salient consumer values with previous buyers does not logically guarantee that a site is trustworthy. Yet our data indicates that salient consumer value similarity did in fact increase trust, suggesting that the SVS model of trust is a valid process model in cases of financial transactions under conditions of social uncertainty.

While we widened the definition of trust by including indicators of store trustworthiness, benevolence, integrity, and competence, ANOVA tests showed that salient consumer value similarity to previous buyers had effects on all four indicators (tables not shown), raising the possibility that people depend on information about salient consumer value similarity to previous buyers to make comprehensive evaluations in situations where they lack adequate information about a website's trustworthiness. In other words, rather than arriving at trust as the sum of independent evaluations of competence, benevolence, and so on, people form top-down judgments based on external information about salient consumer value similarity to previous buyers. This interpretation is supported by the extremely high goodness of fit of the models in Figs. 2.3 and 2.4.

Several tasks remain before we can apply the findings of this study to create a trust-building methodology for e-commerce stores. First is the question of how to present potential buyers with information about SVSs to previous buyers. If previous buyer profiles are displayed on a site, the information becomes an internal component of the website. In fact, reputation information is already widely presented on e-commerce stores in the form of user ratings and other messages from previous buyers. However, such displays may not function as valid clues for potential buyers if they do not discuss salient consumer values or if it is possible for the website owner to manipulate the information. These kinds of internal components may also be insufficient as signals of trustworthiness because, as discussed earlier, they may be easily imitated and reproduced. In this study, information on previous buyers is presented as the result of a survey by a research company independent of the e-commerce store. This retains the neutrality of the external information and suggests that information on salient consumer values and attributes of previous buyers should be presented by a neutral third party.

Of course, not all potential buyers will be willing to trust an e-commerce store just because a neutral third party has provided information about previous buyers. Our study shows that trust building is rather difficult if potential buyers and previous buyers have different salient consumer values. This means that a trust-building methodology based only on the presentation of salient consumer values is not universally effective. However, when the customers of an e-commerce store share some consumer values, presenting such information should be effective in attracting, or selectively encircling, potential buyers who share such values. By explicitly stating that users of a store share the same salient consumer values, e-commerce stores can build a sense of community or in-group identity among users to maintain a highly loyal customer base based on trust.

Finally, our study leaves us with several research tasks for the future. While we tested a food product e-commerce store, it is unclear whether we would obtain the same results with other products or services. Transaction uncertainties are greater for food products because there tends to be greater variation in quality, unlike products such as books or CDs. Further research should be conducted on the effects of similarity on trust in e-commerce in goods and services that have comparatively low uncertainty levels. Furthermore, researchers must consider formats other than survey research results that may be used to present similarities of previous buyers to potential consumers. In reality, potential buyers may judge similarities based on external information posted independently by previous buyers on social media. The difference between the effects of this kind of user-generated content and presentation by third party survey results should also be further explored. By continuing this line of research, we hope to improve understanding of e-commerce trust-building methodologies for both practitioners and researchers.

Appendix

Scale Items

Perceived Store Trustworthiness

(1. Agree, 2. Somewhat agree, 3. Somewhat disagree, 4. Disagree)

1. This store is trustworthy.
2. This store wants to be known as one that keeps its promises and commitments.
3. I trust this store to keep my best interests in mind.
4. I find it necessary to be cautious with this store.
5. The retailer of this store has more to lose than to gain by not delivering on its promises.
6. This store's behavior meets my expectations.

Perceived Benevolence

(1. Agree, 2. Somewhat agree, 3. Somewhat disagree, 4. Disagree)

1. I believe that this store would act in my best interest.
2. If I required help, this store would do its best to help me.
3. This store is interested in my well-being, not just its own.

Perceived Integrity

(1. Agree, 2. Somewhat agree, 3. Somewhat disagree, 4. Disagree)

1. This store is truthful in its dealings with me.
2. I would characterize this store as honest.
3. This store would keep its commitments.
4. This store is sincere and genuine.

Perceived Competence

(1. Agree, 2. Somewhat agree, 3. Somewhat disagree, 4. Disagree)

1. This store is competent and effective in selling crabs.
2. This store performs its role of selling crabs very well.
3. Overall, this store is a capable and proficient Internet crab seller.
4. In general, this store is very knowledgeable about crabs.

Intention to Buy

(1. Agree, 2. Somewhat agree, 3. Somewhat disagree, 4. Disagree)

1. Given the chance, I would consider buying crabs from this store in the future. [Item 1]
2. I probably would not buy from this store. [Item 2]
3. It is likely I would consider purchasing from this store. [Item 3]
4. It is unlikely I would return to this store before making a purchase decision. [Item 4]

References

Cheskin Research (2000) Trust in the wired Americas. www.debmcdonald.com/trust.pdf. Accessed 22 Sept 2010

Cheskin Research and Studio Archetype/Sapient (1999) eCommerce trust study. www.cheskin.com/cms/files/i/articles//17__report-eComm%20Trust1999.pdf. Accessed 22 Sept 2010

Cook J, Wall T (1980) New work attitude measures of trust, organizational commitment, and personal need nonfulfillment. J Occup Psychol 53(2):39–52

Corbitt BJ, Thanasankit T, Yi H (2003) Trust and e-commerce: a study of consumer perceptions. Electron Commer Res Appl 2(3):203–215

Corritore CL, Kracher B, Wiedenbeck S (2003) On-line trust: concepts, evolving themes, a model. Int J Hum-Comput Stud 58(6):737–758

Cvetkovich G, Nakayachi K (2007) Trust in a high-concern risk controversy: a comparison of three concepts. J Risk Res 10(2):223–237

Davis FD (1986) A technology acceptance model for empirically testing new end-user information systems: theory and results. Unpublished doctoral thesis, MIT Sloan School of Management, Cambridge, MA

Davis FD (1989) Perceived usefulness, perceived ease of use and user acceptance of information technology. MIS Quart 13(3):319–340

Doney PM, Cannon JP (1997) An examination of the nature of trust in buyer–seller relationships. J Mark 61(2):35–51

Driscoll JW (1978) Trust and participation in organizational decision making as predictors of satisfaction. Acad Manage J 21:44–56

Earle TE (2004) Thinking aloud about trust: a protocol analysis of trust in risk management. Risk Anal 24(1):169–183

Earle TE, Cvetkovich G (1995) Social trust: toward a cosmopolitan society. Praeger, Westport

Earle TE, Cvetkovich G (1997) Culture, cosmopolitanism, and risk management. Risk Anal 17 (1):55–65

Fogg BJ (2002) Persuasive technology: using computers to change what we think and do. Morgan Kaufmann, San Francisco

Fogg BJ, Tseng H (1999) The elements of computer credibility. In: Proceedings of the 1999 SIGCHI conference on human factors in computing systems, Pittsburgh, pp 80–87

Fogg BJ, Marshall J, Laraki O et al. (2001) What makes web sites credible? A report on a large quantitative study. In: Proceedings of the SIGCHI conference on human factors in computing systems, Seattle, pp 61–68

Fogg BJ, Soohoo C, Danielson DR, Marable L, Stanford J, Tauber ER (2003) How do users evaluate the credibility of websites? A study with over 2,500 subjects. In: Proceedings of the 2003 conference on designing for user experiences, San Francisco, pp 1–15

Gambetta D (1988) Can we trust trust? In: Gambetta D (ed) Trust: making and breaking cooperative relations. Basil Blackwell, New York, pp 213–237

Gefen D (2000) E-commerce: the role of familiarity and trust. Omega 28(6):725–737

Gefen D, Straub DW (2004) Consumer trust in B2C e-commerce and the importance of social presence: experiments in e-products and e-services. Int J Manage Sci 32(6):407–424

Gefen D, Karahanna E, Straub DW (2003a) Inexperience and experience with online stores: the importance of TAM and trust. IEEE Trans Eng Manage 50(3):307–321

Gefen D, Karahanna E, Straub DW (2003b) Trust and TAM in online shopping: an integrated model. MIS Quart 27(1):51–90

Grabner-Kräuter S, Kaluscha EA (2003) Empirical research in on-line trust: a review and critical assessment. Int J Hum-Comput Stud 58:783–812

Hoffman DL, Novak TP, Peralta M (1999) Building consumer trust online. Commun ACM 42 (4):80–85

Hovland CI, Weiss W (1951) The influence of source credibility on communication effectiveness. Public Opin Quart 15:635–650

Hovland CI, Janis IL, Kelly HH (1953) Communication and persuasion. Yale University Press, New Haven

Jarvenpaa SL, Tractinsky N, Saarinen L, Vitale M (1999) Consumer trust in an internet store: a cross-cultural validation. J Comput-Mediat Commun 5(2). http://jcmc.indiana.edu/vol5/issue2/jarvenpaa.html. Accessed 22 Sept 2010

Jarvenpaa SL, Tractinsky N, Vitale M (2000) Consumer trust in an internet store. Inf Technol Manage 1(1–2):45–71

Kee HW, Knox RE (1970) Conceptual and methodological considerations in the study of trust. J Conflict Resol 14:357–366

Kollock P (1999) The production of trust in online markets. Adv Gr Process 16:99–123

Koufaris M, Hampton-Sosa W (2004) The development of initial trust in an online company by new customers. Inf Manage 41(3):377–397

Kover SE, Burke KG, Kover BR (2000a) Consumer response to the CPA WebTrust assurance. J Inf Syst 14(1):17–35

Kover SE, Burke KG, Kover BR (2000b) Selling WebTrust: an exploratory examination of factors influencing consumers' decisions to use online distribution channels. Rev Accoun Inf Syst 4 (2):39–52

Lim KH, Sia CL, Lee MKO, Benbasat I (2001) How do I trust you online, and if so, will I buy? An empirical study on designing web contents to develop online trust. Working paper, University of British Columbia

Lim KH, Sia CL, Lee MKO, Benbasat I (2006) Do I trust you online, and if so, will I buy? An empirical study of two trust-building strategies. J Manage Inf Syst 23:233–266

Lowry PB, Vance A, Moody G, Beckman B, Read A (2008) Explaining and predicting the impact of branding alliances and web site quality on initial consumer trust of e-commerce web sites. J Manage 24(4):199–224

Mayer RC, Davis JH, Schoorman FD (1995) An integrative model of organizational trust. Acad Manage Rev 20:709–734

McGinnies E, Ward CD (1980) Better liked than right: trustworthiness and expertise as factors in credibility. Pers Soc Psychol Bull 6(3):467–472

McKnight DH (2001) What trust means in e-commerce customer relationships: an interdisciplinary conceptual typology. Int J Electron Commer 6(2):35–59

McKnight DH, Chervany NL (2001) What trust means in e-commerce customer relationships: an interdisciplinary conceptual typology. Int J Electron Commer 6(2):35–59

McKnight DH, Chervany NL (2002) Conceptualizing trust: a typology and e-commerce customer relationships model. In: Proceedings of the 34th annual Hawaii international conference on system science (HICSS-34), Maui. www.hicss.hawaii.edu/HICSS_34/PDFs/INCRM04.pdf. Accessed 22 Sept 2010

McKnight DH, Cummings LL, Chervany NL (1998) Initial trust formation in new organizational relationships. Acad Manage Rev 23(3):473–490

McKnight DH, Choudhury V, Kacmar C (2002) Developing and validating trust measures for e-commerce: an integrative typology. Inf Syst Res 13(3):334–359

Pavlou PA (2003) Consumer acceptance of electronic commerce: integrating trust and risk with the technology acceptance model. Int J Electron Commer 7(3):69–103

Pavlou PA, Gefen D (2004) Building effective online marketplaces with institution-based trust. Inf Syst Res 15(1):37–59

Poortinga W, Pidgeon NF (2006) Prior attitudes, salient value similarity and dimensionality: towards an integrative model of trust in risk regulation. J Appl Soc Psychol 36:1673–1699

Reichheld FF, Schefter P (2000) "E-loyalty": your secret weapon on the web. Harv Bus Rev 78 (4):105–113

Ross L, Greene D, House P (1977) The false consensus effect: an egocentric bias in social perception and attribution process. J Exp Soc Psychol 13(3):297–301

Salam AF, Iyer L, Palvia P, Singh R (2005) Trust in e-commerce. Commun ACM 48(2):72–77

Scott CL III (1980) Interpersonal trust: a comparison of attitudinal and situational factors. Hum Relation 33:805–812

Shapiro SP (1987) The social control of impersonal trust. Am J Sociol 93(3):623–658

Siegrist M, Cvetkovich GT (2000) Perception of hazards: the role of social trust and knowledge. Risk Anal 20:713–720

Siegrist M, Cvetkovich GT, Gutscher H (2001) Shared values, social trust, and the perception of geographic cancer clusters. Risk Anal 21:1047–1053

Siegrist M, Earle TC, Gutscher H (2003) Test of a trust and confidence model in the applied context of electromagnetic field (EMF) risks. Risk Anal 23:705–716

Siegrist M, Gutscher H, Earle TC (2005) Perception of risk: the influence of general trust, and general confidence. J Risk Res 8(2):145–156

Stewart KJ (2003) Trust transfer on the world wide web. Organ Sci 14(1):5–17

Wang H, Lee MKO, Wang C (1998) Consumer privacy concerns about internet marketing. Commun ACM 41(3):63–70

Westland JC (2010) Lower bounds on sample size in structural equation modeling. Electron Commer Res Appl 9(6):476–487

Williamson OE (1993) Calculativeness, trust, and economic organization. J Law Econ 34:453–502

Zucker LG (1986) Production of trust: institutional sources of economic structure, 1840–1920. Res Organ Behav 8:53–111

Chapter 3
Information Diffusion and Dissipative Effect on Social Networks

Yuya Dan

3.1 Introduction

Social network analysis (Newman et al. 2006) has become quite important because we can not only visualize the structure of online social networks but also perform a large-scale computation of the phenomena. It has become normal in recent years to apply the theory of social network analysis to IT enabled Services (ITeS). In this chapter, we investigate the process of information diffusion and dissipative effect on social networks through computer simulation (Albert and Barabasi 2002).

Information and Communication Technology (ICT) enables users to share a variety of information or knowledge anywhere and anytime via the Internet. Internet-based social networks, such as Bulletin Board System (BBS), Social Networking Service (SNS), Blogosphere, have in recent years become widely used as repositories of personal information. They have transformed the means of information storage, media, and access (Vazquez et al. 2002).

The speed of information diffusion has become so rapid that we can send messages to others in several seconds compared to several days or even weeks in past. Nowadays we can know events and incidents without resorting to mass media such as Television (TV), radio, newspapers, magazines, and so on. Moreover, people can not only receive but submit an article to share information on social networks.

On the other hand, we are faced by certain difficulties in the confidentiality of sensitive information, which could become widely accessible through using the Internet. Once such information diffuses on the Internet, all users can potentially learn of the information anywhere and anytime.

Social network analysis has become an important research topic in mathematical analysis and modeling for its application for social media marketing (Dellarocas

Y. Dan (✉)
Faculty of Business Administration, Matsuyama University, Matsuyama, Ehime, Japan
Mathematisches Institut, Ludwig-Maximilians-Universität, München, Deutschland
e-mail: dan@cc.matsuyama-u.ac.jp

S. Uesugi (ed.), *IT Enabled Services*,
DOI 10.1007/978-3-7091-1425-4_3, © Springer-Verlag Wien 2013

2003; Leskovec et al. 2007), election strategy for voters' attitudes (Huckfeldt and Sprague 1991), and information security (Nikoloski et al. 2006; Quing and Wen 2005).

It is interesting and important to analyze the process of information diffusion and dissipative effect on social networks. The phenomena of information diffusion is a random process that creates a complex system of interacting users mainly over the Internet. In particular, social networks on the Internet are considered to be scale-free networks which have power-law distribution in degree of links. Understanding the dynamics of information diffusion and dissipative effect on social networks is fundamental and therefore the first step towards devising effective techniques in all the fields of ITeS (Aral et al. 2007).

This chapter is organized as follows: We see the present state of social networks on the Internet in Sect. 3.2. According to the nature of social networks as information media, we point out the risk of diffusion of personal information and dissipative effect on social networks in Sect. 3.4. Then we make mathematical modeling of the process of information diffusion and dissipative effect in Sect. 3.5. We discuss mathematical analysis of diffusion phenomena on social networks in Sect. 3.6. As the main results of this chapter, simulations of information diffusion and dissipative effect on social networks are described in Sect. 3.7. Finally, we conclude our works in the last section.

3.2 Social Networks

The present state of social networks on the Internet is described in this section.

The nature of *small world* in social networks was discovered by Milgram (1967) in his psychological experiment. In the experiment, a sample set of individuals in Kansas were asked to reach a particular target person in Boston by passing a message in a letter with their names and addresses along a chain of acquaintances. The average length of successful chains turned out to be five intermediaries or six separation steps in the social network in the United States.

A recent online *small world* experiment at Columbia University found that five to seven degrees of separation is sufficient for connecting any two people through e-mail networks (Watts 2003). See the work by Watts and Strogatz (1998) for more information about small world more in detail.

3.2.1 SNS

SNS is one of the most famous services on the Internet.

A SNS is a platform of online activities. The users in the SNS can invite other friends, share their profile, submit articles and photos, exchange messages, enjoy social games and so on. SNSs are constructed by the relationships among users.

Table 3.1 Examples of SNSs

Service name	Start	Registered members (million)	Network structure
Facebook	Feb. 2004	800	Bidirectional link
Google+	Jun. 2011	25	Unidirectional link
GREE	Feb. 2004	25	Unidirectional link
mixi	Feb. 2004	40	Bidirectional link
LinkedIn	May 2003	100	Bidirectional link
Mobage	Feb. 2006	20	Unidirectional link
MySpace	Aug. 2003	200	Bidirectional link

For instance, Facebook[1] has over 800 million registered users; that is larger than the population of any country except for China and India. Facebook users can join with their name registration, e-mail address and other profile information, and can then search for, and make, friends on Facebook. These relationships can be viewed by other friends, thereby developing 'friend of friend' chains on Facebook.

Thus the network can be described as being constructed and evolved according to the real relationships among friends.

See other SNSs in Table 3.1: Google + ,[2] GREE,[3] mixi,[4] LinkedIn,[5] Mobage,[6] MySpace,[7] are all famous online services.

SNS seems to be a visualized relationship of friends from our real lives. On the other hand, anyone can anonymously access this personal information. Once such a piece of personal information diffuses on the SNS, nobody can control the flow of information. We should become prudent when we use personal, confidential and potentially harmful information on SNSs.

3.2.2 Blogosphere

The blogosphere can be defined as including all blogs on the World Wide Web (WWW). Typically, an individual blogger writes an article on his/her blog, and after having read the article, another blogger gives a comment or trackback to the article. The information diffuses according to the hyperlinks on the WWW (Table 3.2).

---

[1] http://www.facebook.com/

[2] http://plus.google.com/

[3] http://gree.jp/

[4] http://mixi.jp/

[5] http://www.linkedin.com/

[6] http://mbga.jp/

[7] http://www.myspace.com/

Table 3.2 Examples of blog services

Service name	Start	Registered members (million)	Network structure
Ameba BLOG	Sep. 2004	200	Unidirectional link
Google Buzz	Feb. 2010	Unknown	Unidirectional link
Twitter	Jun. 2006	300	Unidirectional link
Tumblr	Mar. 2007	40	Unidirectional link

The algorithm of PageRank proposed by Page et al. (1999) is an important tool of hyperlink analysis. PageRank gives the importance of each Web page using hyperlink structure on the WWW.

Information itself is neither good nor bad, it is how the information is contextualized that gives it this quality. Accordingly social networks can be used to circulate information; fake, true, maliciously, or unwittingly. Blog under fire or blog flaming is a phenomena whereby the readers of a blog write a comment to a blog article and 'blow up' the number of comments contrary to the blogger's expectation (Table 3.2).

3.2.3 E-mail

Electronic mail (e-mail) service is popular on the Internet. E-mail enables users to send messages comprising of one, or a combination or; images, videos, or other data files as well as text messages via e-mail client software. Almost all users of e-mail client software make their lists of e-mail addresses of friends or acquaintances. This relationship in the e-mail service is also said to be a social network.

There are e-mail worms in malware which spread on social networks constructed by address books in the e-mail client software. Infection and the spread of e-mail worms represent a key factor in the field of information security.

A diffusion model of worms on scale-free networks is proposed by Nikoloski et al. (2006; Quing and Wen 2005) using differential equations.

3.3 Diffusion of Confidential Information

3.3.1 Risk of Diffusion

As is described in the previous sections, we focused on the diffusion of personal or secret information that are quite important in digital societies. Once confidential information is known widely on the networks, methods do not currently exist by which the situation can be contained.

It is important to discuss the diffusion of personal information on networks from the point of information security. According to the result of the simulation, we can conclude that the diffusion rate on scale-free networks is the fastest among current networks. This conclusion means that the leak of private information or pictures causes serious problem on the online social networks (Dan 2011a, b).

User Generated Content (UGC) is currently popular for exchanging information amongst Internet users. Orita (2008) pointed out that users in Japan prefer to remain anonymous although they make good use of such UGC. Once the personal information of users is open on the Internet, it diffuses on the network, all users in the community can access the information, and it becomes too difficult to go back to the state of anonymity.

Even if we include the effect of deletion, which is linear in time, we could not stop the diffusion of personal information throught social networks on the Internet. As is shown in the simulate on later, the diffusion of information progresses at the rate of exponential growth. Thus, deletion is ineffective as we completely delete all references to the information in digital format.

3.3.2 Dissipative Effect

There are at least two approaches to dissipative operations concerning the information diffusion process. One is to eliminate the main hub in a scale-free network, the other is to decrease the diffusion speed. We can realize the latter operation through the reduction of the probability to diffuse information on the network. It is proposed in this chapter that the dissipative operations are effective in protection of confidential information.

For the protection of personal information, it should be pointed out that there are three other options. For technical reasons, we should use cryptography to store and communicate personal information. Secondly, from the viewpoint of the law system, we should develop and manage appropriate measures concerning the protection of personal information. Lastly but most importantly, we need to educate or enlighten users.

3.4 Mathematical Modeling

3.4.1 Set Theoretic Model

A social network (Newman 2010) can be modeled using mathematics, that is, a set of points (also called *vertices* or *nodes*) connected by lines (also called *edges* or *links*).

Let us consider a social network according to the usual method by Wasserman and Faust (1994). U is defined below as a set of users in the social network

$$U = \{u_1, u_2, \ldots, u_n\}, \tag{3.1}$$

where n is the number of users in the social network. The elements of U are points of a graph in the geometrical view.

We can represent the connection among users in social networks as a subset of $U \times U$. We define $R \subset U \times U$, using $(u_i, u_j) \in R$ if $u_i \in U$ is connected to $u_j \in U$. For example, we see that $U = \{u_1, u_2, u_3\}$ and $R = \{(u_2, u_1), (u_3, u_1), (u_3, u_2)\}$ describes a social network consists of three users where u_2 is connected to u_1, and u_3 is connected to u_1 and u_2. It should be remarked that (u, u) could not belong to R for any $u \in U$.

We say a graph is symmetric when the relations between two users are directional, that is, $(u_i, u_j) \in R$ imples $(u_j, u_i) \in R$ for every element of R. Both Facebook and mixi are symmetric in the relation of social networks, however, Twitter and e-mail are not symmetric.

3.4.2 Adjacency Matrix

As another representation of the connection of social networks, we can use an adjacency matrix. The elements of an adjacency matrix for the relation R

$$
A = \begin{pmatrix}
a_{11} & a_{12} & \cdots & a_{1n} \\
a_{21} & \ddots & & \vdots \\
\vdots & & \ddots & \vdots \\
a_{n1} & \cdots & \cdots & a_{nn}
\end{pmatrix}
\tag{3.2}
$$

is defined without loss of generality by

$$
a_{ij} = \begin{cases} 1 \text{ if } (u_i, u_j) \in R \\ 0 \text{ if } (u_i, u_j) \notin R. \end{cases}
\tag{3.3}
$$

for every $i, j = 1, 2, \ldots, n$.

For example, if we take the same social network U and R in the previous section, then the adjacency matrix is expressed by

$$
A = \begin{pmatrix} 0 & 0 & 0 \\ 1 & 0 & 0 \\ 1 & 1 & 0 \end{pmatrix}.
\tag{3.4}
$$

3.4.3 Markov Chain Model

The Markov chain model is introduced as a stochastic process in the phenomena of information diffusion in discrete time steps.

For a social network U with R, initial data of a state of the users can be written by

$$
v = (v_1, v_2, \ldots, v_n) \in \{0, 1\}^n,
\tag{3.5}
$$

where n is the number of users in the social network. We can have the next state of users as follows:

$$vA = (v_1, v_2, \ldots, v_n) \begin{pmatrix} a_{11} & a_{12} & \cdots & a_{1n} \\ a_{21} & \ddots & & \vdots \\ \vdots & & \ddots & \vdots \\ a_{n1} & \cdots & \cdots & a_{nn} \end{pmatrix}. \tag{3.6}$$

Hence, we can obtain m-time steps later,

$$vA^m = (v_1, v_2, \ldots, v_n) \begin{pmatrix} a_{11} & a_{12} & \cdots & a_{1n} \\ a_{21} & \ddots & & \vdots \\ \vdots & & \ddots & \vdots \\ a_{n1} & \cdots & \cdots & a_{nn} \end{pmatrix}^m \tag{3.7}$$

as the result of the Markov chain model. We use the logical sum instead of the algebraic sum in our simulation for the memory effect.

3.5 Analysis

3.5.1 Logistic Curves

Rogers (2003) has investigated a wide variety of diffusion phenomena. Figure 3.1 shows logistic curves

$$f_a(t) = \frac{1}{1 + e^{-at}} \tag{3.8}$$

introduced by Verhulst (1838) with different continuous parameter $a \in [0, 2]$. When $f_a(t)$ means the probability to reach a particular information at time t, the function (3.8) describes the percolation of the information.

In fact, $f_a(t)$ satisfies the ordinary differential equation

$$\frac{d}{dt} f_a(t) = a f_a(t) \left(1 - f_a(t)\right). \tag{3.9}$$

This means the increase rate or slope of $f_a(t)$ is proportional both to $f_a(t)$ and $1 - f_a(t)$. It is clear that $f_a(t)$ becomes flat far from the origin, where $f_a(t)$ takes the value near 0 or 1. It also should be remarked that $f_a(t)$ increases the most rapidly at $t = 0$, where $f_a(t)$ takes the value of 1/2. We know the quadratic form $x(1 - x)$ takes the value between 0 and 1/4.

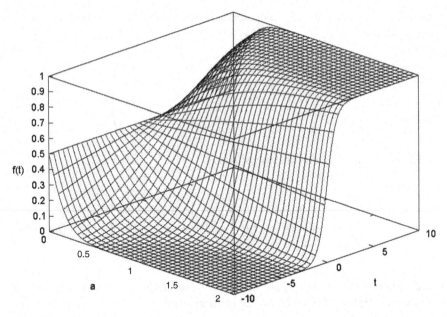

Fig. 3.1 Logistic curves $f_a(t)$ with different continuous parameter $a \in [0, 2]$. We can see that $f_a(t)$ remains be a constant if $a = 0$, and that $f_a(t)$ grows more rapid near $x = 0$ if a is larger

These discussions are quite important because we can know the phenomena not only from the form of solutions to differential equations, but also from the differential equations themselves.

3.5.2 Diffusion in Continuous Spaces

When we discuss the diffusion on flat continuous spaces, Euclidean spaces are considered to be fundamental metric spaces. Therefore, we can apply the theorems in geometry and analysis.

Let us consider the diffusion phenomena on Euclidean space $\mathbb{R}^n$. The density in the diffusion phenomena is a function of time and space, so that we can write

$$u : (t, x) \in \mathbb{R} \times \mathbb{R}^n \rightarrow \mathbb{R}. \tag{3.10}$$

In general, the diffusion coefficients for density is not uniform in $\mathbb{R}^n$, that is, the diffusion speeds are different for each point in the space. The diffusion phenomena obeys the diffusion equation

$$\frac{\partial}{\partial t} u(t, x) = \nabla \cdot (D(x) \nabla u(t, x)), \tag{3.11}$$

where $D(x)$ denotes the diffusion coefficients which depend on the structure of space, $\partial/\partial t$ the differential operator with respect to the time variable and

$$\nabla = \left(\frac{\partial}{\partial x_1}, \frac{\partial}{\partial x_2}, \ldots, \frac{\partial}{\partial x_n}\right) \tag{3.12}$$

the differential operator with respect to the space variables. Here we use the inner product

$$a \cdot b = a_1 b_1 + a_2 b_2 + \cdots + a_n b_n \tag{3.13}$$

with two vectors, say $a = (a_1, a_2, \ldots, a_n) \in \mathbb{R}^n$ and $b = (b_1, b_2, \ldots, b_n) \in \mathbb{R}^n$. The diffusion equation (3.11) is a partial differential equation (PDE) of time and space with nonlinear forms.

If there is a positive constant D_0 such that $D(x) = D_0$ for any $x \in \mathbb{R}^n$, then we have the linear form

$$\frac{\partial}{\partial t} u(t, x) = D_0 \Delta u(t, x), \tag{3.14}$$

of a diffusion equation, where $\Delta = \nabla \cdot \nabla$ is the Laplace operator in $\mathbb{R}^n$.

Fourier analysis tells us the solution to (3.14) with initial data $u(0, x) = u_0(x)$ for any $x \in \mathbb{R}^n$,

$$u(t, x) = (2\pi)^{-\frac{n}{2}} \int_{\mathbb{R}^n} e^{ix \cdot \xi} e^{-t\frac{|\xi|^2}{4D_0}} \hat{u}_0(\xi)\, d\xi, \tag{3.15}$$

where $\hat{u}_0$ stands for the Fourier transform of u_0 with respect to space variables

$$\hat{u}_0(\xi) = (2\pi)^{-\frac{n}{2}} \int_{\mathbb{R}^n} e^{-ix \cdot \xi} u_0(x)\, dx. \tag{3.16}$$

It is also possible to define calculus on networks. See Chap. 2 in Cardanobile (2010) for integration by parts on networks.

3.5.3 Diffusion on Networks

In this section, we consider the diffusion phenomena of information on the variety of networks, such as complete, random, stochastic and scale-free networks. We provide each definition of the corresponding models of networks. The dynamics of diffusion or percolation depends on the structure of networks. We see the property of networks under the definition, and consider the characteristics of each network. These results are a natural generalization of the previous works by Dan (2011a).

3.5.3.1 Complete Networks

A complete network is the network of whose two vertices have an edge. There is no pair that does not have an edge in the network. When the number of vertices is n, the network has $n(n - 1)/2$ edges. In our notation, a social network U with $R = U \times U$ is a complete network. Dan (2011a) investigated the mathematical modeling and computer simulation of diffusion phenomena on social networks for complete networks.

3.5.3.2 Random Networks

A random network is one whose vertices have edges at random. Randomness is assumed for not only uniform distribution, but also any possible function of distribution. Dan (2011b) has pointed out that the structure of random networks is similar to that of stochastic networks in diffusion processes.

3.5.3.3 Stochastic Networks

Conversely, for a stochastic network, each edge has a probability value between zero and one. Each edge mediates the information at the probability that depends on the edge. One can communicate on the edge at the probability p, otherwise one cannot communicate on the edge at the probability $1 - p$. The possibility of communication depends on the probability p defined for each edge. In the simulation, we use uniform stochastic networks with a constant p for dissipative operations.

3.5.3.4 Scale-Free Networks

In a scale-free network, the power-law is used for the number of edges. There are some vertices, which are called *hubs*, that have comparably large number of edges. On the other hand, almost all vertices have only a few edges. The graph of the number of edges indicates the law of power. Scale-free networks were first proposed as small-world networks by Watts and Strogatz (1998), then Barabási and Albert (1999) have constructed their models for scale-free networks.

It is known that scale-free networks have high cluster coefficients like regular lattices. However, these networks have small characteristic path lengths like random networks.

We use scale-free networks in our simulation.

Table 3.3 Simulation environment

Item	Components and performance
CPU	Intel Pentium D (2MB × 2 L2 Cache, 3.20 GHz, 800 MHz FSB)
Memory	3 GB DDR (DDR2-SDRAM, Dual Channel)
OS	Linux 2.6.33.3-85.fc13.c86 64
Compiler	gcc 4.4.4
Option	Default optimization

3.6   Simulation

In this section, we will investigate the information diffusion on scale-free networks that often appear in our realistic social networks. According to our model of information diffusion, we can see the simulation of phenomena using computational calculations.

3.6.1   Settings

In our simulation, the hardware comprised an Intel Pentium D with 64 bit mode and 3 GB DDR. Our simulation program was a native C program, which was compiled by gcc 4.4.4 on Fedora 13 (Linux 2.6.33.3-85.fc13.c86 64) with default optimization. See Table 3.3 for components in detail.

At the beginning of the simulation, the program constructs a scale-free network according to the method of preference selection proposed by Dorogovtsev et al. (2000). Figure 3.2 shows the degree distribution of the generated network. In the figure, the frequency of each degree, which is defined by the number of links of a node, are plotted with log-log axis. It is easy to see from the figure that the degree distribution forms like

$$p(k) = Ck^{-\gamma}, \tag{3.17}$$

where k is the degree of each node, γ is the scale factor and C is a positive constant. We have obtained $\gamma = 1.77$ and $C = 333$ by applying regression with determination coefficient $R^2 = 0.904$.

In the simulation,

$$U = \{u_1, u_2, \ldots, u_{1024}\} \tag{3.18}$$

is assumed because of the memory limitation of two-dimensional array alignments. A part of the adjacency matrix is expressed by

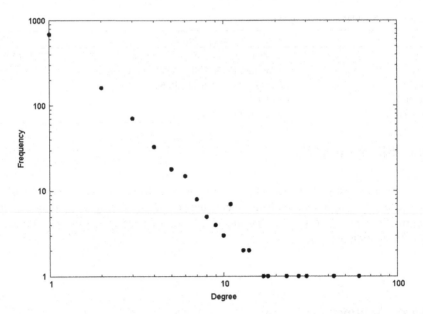

Fig. 3.2 The degree distribution of the network which is used in our simulation. We can read from the plot on the *right-hand side* that there is a node which has 60 links, a node which has 43 links, and so on. On the *left-hand side* there are 687 nodes which have only 1 link without connections to others

$$
\begin{pmatrix}
1 & 1 & 1 & 1 & 0 & 0 & 1 & 0 & 0 & 0 & 0 & 1 & 0 & 0 & 0 & 0 & 1 & 0 & 1 & 0 & \cdots \\
1 & 0 & 0 & 0 & 1 & 1 & 0 & 0 & 0 & 0 & 1 & 0 & 0 & 0 & 1 & 0 & 0 & 1 & 0 & 1 & \cdots \\
1 & 0 & 0 & 0 & 0 & 0 & 0 & 0 & 0 & 0 & 0 & 0 & 0 & 0 & 0 & 0 & 0 & 0 & 0 & 0 & \cdots \\
1 & 0 & 0 & 0 & 0 & 0 & 0 & 0 & 0 & 0 & 0 & 0 & 0 & 0 & 0 & 0 & 0 & 0 & 0 & 0 & \cdots \\
0 & 1 & 0 & 0 & 0 & 0 & 0 & 0 & 0 & 0 & 1 & 0 & 0 & 0 & 0 & 0 & 0 & 0 & 0 & 0 & \cdots \\
0 & 1 & 0 & 0 & 0 & 0 & 0 & 1 & 1 & 0 & 0 & 0 & 1 & 1 & 0 & 0 & 0 & 0 & 0 & 0 & \cdots \\
1 & 0 & 0 & 0 & 0 & 0 & 0 & 0 & 0 & 0 & 0 & 0 & 0 & 0 & 0 & 0 & 0 & 0 & 0 & 0 & \cdots \\
0 & 0 & 0 & 0 & 0 & 1 & 0 & 0 & 0 & 0 & 0 & 0 & 0 & 0 & 0 & 0 & 0 & 0 & 0 & 0 & \cdots \\
0 & 0 & 0 & 0 & 0 & 1 & 0 & 0 & 0 & 0 & 0 & 0 & 0 & 0 & 0 & 0 & 0 & 0 & 0 & 0 & \cdots \\
0 & 0 & 0 & 0 & 1 & 0 & 0 & 0 & 0 & 0 & 0 & 0 & 0 & 0 & 0 & 0 & 0 & 0 & 0 & 0 & \cdots \\
0 & 1 & 0 & 0 & 0 & 0 & 0 & 0 & 0 & 0 & 0 & 0 & 0 & 0 & 0 & 0 & 0 & 0 & 0 & 0 & \cdots \\
1 & 0 & 0 & 0 & 0 & 0 & 0 & 0 & 0 & 0 & 0 & 0 & 0 & 0 & 0 & 0 & 0 & 0 & 0 & 0 & \cdots \\
0 & 0 & 0 & 0 & 0 & 1 & 0 & 0 & 0 & 0 & 0 & 0 & 0 & 0 & 0 & 0 & 0 & 0 & 0 & 0 & \cdots \\
0 & 0 & 0 & 0 & 0 & 1 & 0 & 0 & 0 & 0 & 0 & 0 & 0 & 0 & 0 & 1 & 0 & 0 & 0 & 0 & \cdots \\
0 & 1 & 0 & 0 & 0 & 0 & 0 & 0 & 0 & 0 & 0 & 0 & 0 & 0 & 0 & 0 & 0 & 0 & 0 & 0 & \cdots \\
0 & 0 & 0 & 0 & 0 & 0 & 0 & 0 & 0 & 0 & 0 & 0 & 0 & 1 & 0 & 0 & 0 & 0 & 0 & 0 & \cdots \\
1 & 0 & 0 & 0 & 0 & 0 & 0 & 0 & 0 & 0 & 0 & 0 & 0 & 0 & 0 & 0 & 0 & 0 & 0 & 0 & \cdots \\
0 & 1 & 0 & 0 & 0 & 0 & 0 & 0 & 0 & 0 & 0 & 0 & 0 & 0 & 0 & 0 & 0 & 0 & 0 & 0 & \cdots \\
1 & 0 & 0 & 0 & 0 & 0 & 0 & 0 & 0 & 0 & 0 & 0 & 0 & 0 & 0 & 0 & 0 & 0 & 0 & 0 & \cdots \\
0 & 1 & 0 & 0 & 0 & 0 & 0 & 0 & 0 & 0 & 0 & 0 & 0 & 0 & 0 & 0 & 0 & 0 & 0 & 0 & \cdots \\
\vdots & \vdots & \vdots & \vdots & \vdots & \vdots & \vdots & \vdots & \vdots & \vdots & \vdots & \vdots & \vdots & \vdots & \vdots & \vdots & \vdots & \vdots & \vdots & \vdots & \ddots
\end{pmatrix}
$$

$$(3.19)$$

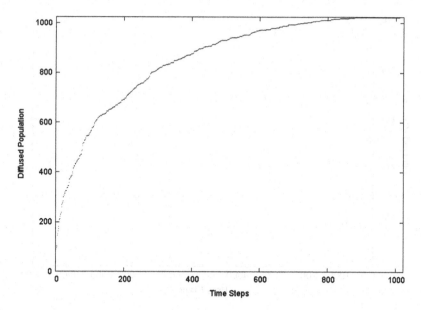

Fig. 3.3 The result of diffusion process in a trial of the simulation. The population who knows the information or are infected increases 60 at the first percolation, up to 89 at the next percolation, and so on. We can see the saturation point, that is, the information is diffused by all people until the 1023rd time step

We can calculate the number of links which the user $u_i \in U$ has, as the following quantity:

$$\sum_{j=1}^{1,024} a_{ij} \tag{3.20}$$

As the nature of preference selection method, u_i is likely to have more links if i is small (Kullmann and Kertész 2001).

Next, we gave the initial condition to the simulation program. Take one user $u_i \in U$ who has known a piece of information or been infected by something at first. Then the users who have the links to u_i can reach the information at the next time step. Therefore the information diffuses through percolation. All we want to know is how many people know a piece of information or are infected at each time step, because we can see the diffusion process and the speed of information diffusion on the networks. The program of the simulation makes a loop percolation as time increased at time steps. Figure 3.3 indicates the result of a trial of the simulation. It can be proved in general that the information diffused to all the people until the number of time steps is equal to the number of elements of U, if the initial user of information diffusion is the main hub of the network.

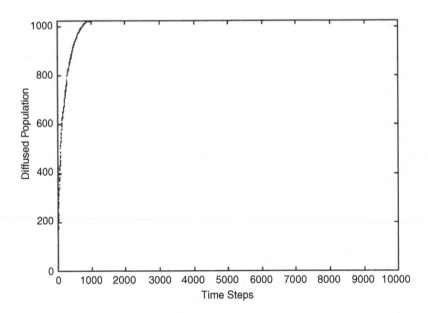

Fig. 3.4 The result of diffusion process with the initial value of u_0

3.6.2 Results for Diffusion

According to the settings of the simulation, we can implement the program under 1,024 possible initial conditions. Since the number of links of u_i tends to decrease as i becomes large, we pick up a typical result of initial value of u_i for $i = 0; 128; 256; 384; 512; 640; 768; 896$ and $1, 024$.

Figure 3.4 shows the case of diffusion process with the initial value of u_0, which is the same initial condition of Fig. 3.3 except for the scaling of axis of time steps. The population grows in the earlier time steps, and reaches the saturation point at $t = 1, 012$.

Figure 3.5 shows the case of diffusion process with the initial value of u_{128}. The population remains 9 or below until the quantum leap of 9–50 at time step 1,030. The quantum leaps also occur at time step 2,050 (23–259) and 3,073 (564–713). The population reaches the saturation point at time step 4,085.

Figure 3.6 shows the case of diffusion process with the initial value of u_{256}. The population remains 1 until the small quantum leap of 1–16 at time step 79. The critical quantum leaps are observed at time step 5,126 (110–149), 6,146 (201–229) and 7,169 (654–713). The population reaches the saturation point at time step 8,181.

Figure 3.7 shows the case of diffusion process with the initial value of u_{384}. The population remains 10 or below until the critical quantum leap of 10–69 at time step 1,025. There is only one quantum leap in the result. The population reach the saturation point at time step 2,037.

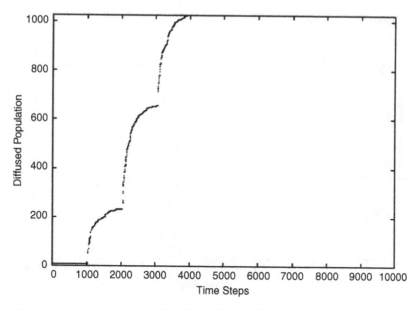

Fig. 3.5 The result of diffusion process with the initial value of u_{128}

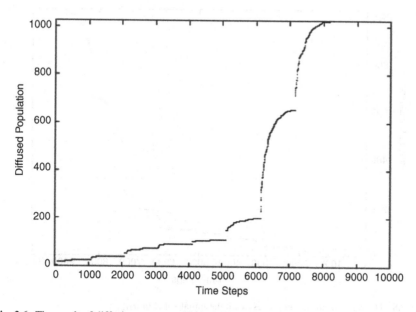

Fig. 3.6 The result of diffusion process with the initial value of u_{256}

Figure 3.8 shows the case of diffusion process with the initial value of u_{512}. The population remains 1 until the growth of 1–3 at time step 299, and remains 7 or below until the small quantum leap of 7–21 at time step 1,103. There are several

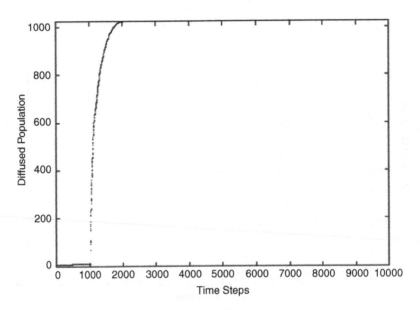

Fig. 3.7 The result of diffusion process with the initial value of u_{384}

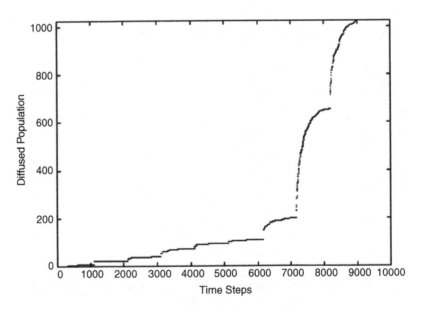

Fig. 3.8 The result of diffusion process with the initial value of u_{512}

quantum leaps of 110–149 at time step 6,150, 201–229 at time step 7,170 and 654–713 at time step 8,193. The population reaches the saturation point at time step 9,205.

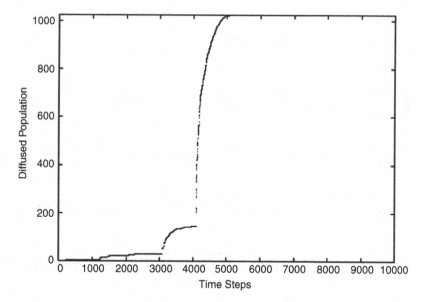

Fig. 3.9 The result of diffusion process with the initial value of u_{768}

Figure 3.9 shows the case of diffusion process with the initial value of u_{768}. The population remains 3 or below until the growth of 3–9 at time step 1,232. There is a quantum leap of 30–53 at time step 3,074 and a critical quantum leap of 114–203 at time step 4,097. The population reaches the saturation point at time step 5,109.

Figure 3.10 shows the case of diffusion process with the initial value of u_{896}. The population remains 7 or below until the growth gets up to 10 at time step 3,179. There is a quantum leap of 40–69 at time step 6,146 or 69–111 at time step 6,150. A quantum leap occurs also at time step 7,169 (694–753). The population reaches the saturation point at time step 8,181.

Figure 3.11 shows the diffusion process with a case whose initial value was u_{1024}. The population remains 3 or below until the growth gets up to 23 at time step 1,026. There is a critical quantum leap of 79–138 at time step 2,049. The population reaches the saturation point at time step 3,061.

3.6.3   Results for Dissipative Effect

Figure 3.12 indicates the result of dissipative effect if the main hub with 60 links is eliminated from the network. Since there no main hub exists, the major quantum leap cannot occur in the simulation. The saturation point of the diffusion population is 694 which cannot exceed the number of diffused users when there is the main hub in the social network.

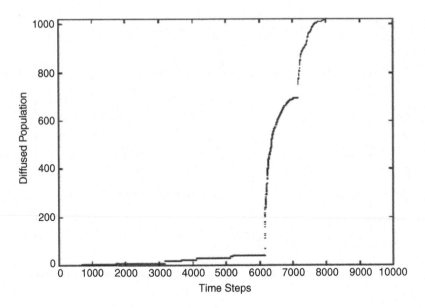

Fig. 3.10 The result of diffusion process with the initial value of u_{896}

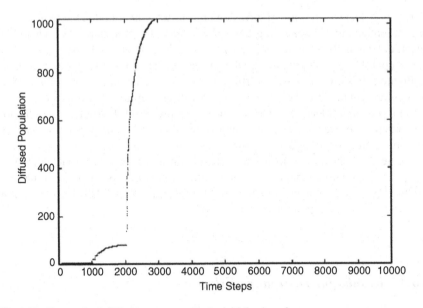

Fig. 3.11 The result of diffusion process with the initial value of u_{1024}

Figure 3.13 shows the result of dissipative effect if the information diffuses probabilistically to other users.

We can conclude that the effect of eliminating the main hub from the network is similar to that of probabilistic diffusion process with the probability valued at around $p = 0.30$ to $p = 0.45$.

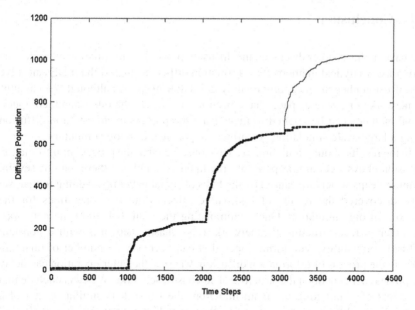

Fig. 3.12 The result of dissipative effect if the main hub of the network is eliminated. The *thin solid line* in the figure denotes the diffusion that is the same as Fig. 3.5 and the *bold dotted line* denotes the diffusion with dissipative effect. It is effective in the quantum leap when the main hub diffuses information to other users

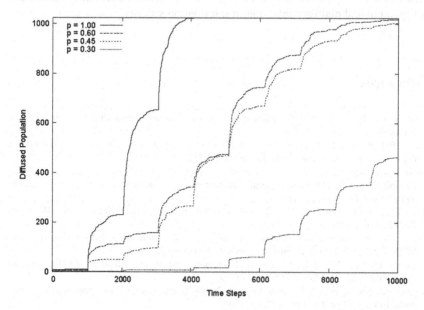

Fig. 3.13 The result of dissipative effect if the information diffuses probabilistically to other users. The results indicates that the information can diffuse with the probabilities to $p = 1.00; 0.60; 0.45; 0.30$ respectively. The lower the probability is, the slower the diffused population grows

3.7 Conclusion

We have in this chapter discussed the diffusion process of information on scale-free networks. Analytical methods for information diffusion suggest that PDE can solve the diffusion phenomena approximately in Euclidean spaces, although the calculus on networks or discrete spaces have been proposed. On the other hand, computational methods can be applied to investigate the process of information diffusion using a large-scale simulation performed on personal desktop computers.

In the results of our simulation, we have observed that the process of information diffusion obeys a certain property of growth curves, and that there may be several quantum leaps which are caused by the hubs of social networks. Adding to that, we have discovered the results of dissipative effect using two operations for the network in our simulation. One is eliminating the main hub from the networks, the other is reconstructing the network as stochastic networks with a constant diffusion probability. The former operation can decrease the number of quantum leaps in the process of information diffusion, whereas the latter operation can delay the process dependent on the structure of stochastic networks. We can conclude that the effect of eliminating the main hub from the network is similar to that of a probabilistic diffusion process with the probability between $p = 0.30$ and $p = 0.45$.

Moreover, our conclusion suggests that we can obtain the results of information diffusion and dissipative effect through further investigation into information management and malware infection in information security.

References

Albert R, Barabási A-L (2002) Statistical mechanics of complex networks. Rev Mod Phys 74 (1):47–97

Aral S, Brynjolfsson E, Alstyne MWV (2007) Productivity effects of information diffusion in networks. MIT Center for Digital Business, Working Paper #234

Barabási A-L, Albert R (1999) Emergence of scaling in random networks. Science 286:509–512

Cardanobile S (2010) Diffusion systems and heat equations on networks. Suedawestdeutscher Verlag Fuer, Hochschulschrif

Dan Y (2011a) Modeling and simulation of diffusion phenomena on social networks. IEEE Proc ICCMS 1:139–146

Dan Y (2011b) Mathematical analysis and simulation of information diffusion on networks. SAINT 2011 Workshop: IT enabled Services (ITeS), pp 550–555

Dellarocas C (2003) The digitization of word of mouth: promise and challenges of online feedback mechanisms. Manage Sci 49(10):1407–1424

Dorogovtsev SN, Mendes JFF, Samukhin AN (2000) Structure of growing networks with preferential linking. Phys Rev Lett 85:4633–4636

Huckfeldt R, Sprague J (1991) Discussant effect on vote choice: intimacy, structure and interdependence. J Polit 53(1):122–158

Kullmann L, Kertész J (2001) Preferencial growth: exact solution of the time dependent distributions. Phys Rev E, 63(051112)

Leskovec J, Adamic L, Huberman BA (2007) The dynamics of viral marketing. ACM Trans Web 1(1), Article No. 5

Milgram S (1967) The small world problem. Psychol Today 1(1):60–67

Newman MEJ (2010) Networks. Oxford University Press, Oxford

Newman MEJ, Barabási A-L, Watts DJ (2006) The structure and dynamics of networks. Princeton University Press, Princeton

Nikoloski Z, Deo N, Kucera L (2006) Correlation model of worm propagation on scale-free networks. Complexus 3:169–182

Orita A (2008) Users attitude towards anonymous and real-name services on the internet in Japan In: JPAIS session at international conference on computer and information science (ICIS), Paris

Page L, Brin S, Motwani R,Winograd T (1999) The PageRank citation ranking: bringing order to the web. Technical Report, Stanford InfoLab

Quing S, Wen W (2005) A survey and trends on internet worms. Comput Secur 24:334–346

Rogers EM (2003) Diffusion of innovations, 5th edn. Free Press, New York

V´azquez A, Pastor-Satorras R,Vespignani A (2002) Large-scale topological and dynamical propertes of the internet. Phys Rev E 65(066130)

Verhulst P-F (1838) Notice sur la loi que la population poursuit dans son accroissement. Correspondance mathématique et physique 10:113–121

Wasserman S, Faust K (1994) Social network analysis: methods and applications. Cambridge University Press, Cambridge

Watts D (2003) Six degrees: the science of a connected age. W.W. Norton, New York

Watts DJ, Strogatz SZ (1998) Collective dynamics of 'small-world' networks. Nature 393:440–442

Chapter 4
Construction of an Appropriately Professional Working Environment for IT Professionals: A Key Element of Quality IT-Enabled Services

Kiyoshi Murata

4.1 Introduction

In industrial nations, information technology (IT) is ubiquitous, and is necessary for a great number of individual and organisational activities. As a result of IT development centred on database and network technology, and the explosive growth of the Internet, a great majority of organisations conduct operations in an e-business environment, wherein most communication is conducted via Internet technology. It is no exaggeration to claim that the quality of people's home, work, and social life in general significantly depends on the quality of IT-enabled services they receive.

A consequence of society's dependence on IT-enabled services is that defects and malfunctions in the services along with IT abuse, cause serious, and sometimes catastrophic, situations. Malicious or negligent development or use of IT-enabled services (or disservices) has led to a number of incidents that have infringed on human rights and corroded human values. Thus, an IT-dependent society is vulnerable. The development and maintenance of high-quality information systems which function as a platform of IT-enabled services, and their reliable operation, are essential to the dependable and secure functioning of society as a whole.

Since business organisations play such a major role in the development and utilisation of IT-based information systems, IT technology professionals who are employed by business organisations, and develop and maintain systems for IT-enabled services, have both intentional and unintentional power over a broad range of people. They must recognise their responsibility to the general public and develop their professional ethics and outlook in order to maintain safety and security, both in the e-business environment and in society.

There are already well-organised, carefully developed codes of IT professional conduct to guide IT professionals in their professional behaviour (e.g. Gotterbarn

K. Murata (✉)
Centre for Business Information Ethics, Meiji University, Tokyo, Japan
e-mail: kmurata@kisc.meiji.ac.jp

S. Uesugi (ed.), *IT Enabled Services*,
DOI 10.1007/978-3-7091-1425-4_4, © Springer-Verlag Wien 2013

et al. 1999). However, these codes of conduct may not function well alone. We cannot ignore the fact that the majority of IT professionals are employed by for-profit businesses. Moreover, any code of conduct is subject to interpretation, and the extent to which IT professionals in business organisations actually follow a code of conduct tends to be influenced by their organisational and social structures and cultures.

This chapter will clarify the conditions that must be met in order to enhance professional outlook and ethical behaviour of IT professionals that will help to provide safe and reliable IT-enabled services:

- IT professionals should develop their sense of professional ethics and their professional attitude; and
- Organisational and social measures should be taken to establish appropriately professional working environments, in which IT professionals are supported in behaving according to their professional code of conduct.

The next section points out how important a sense of professional ethics and professional attitudes are to IT professionals. Section 4.3 examines some typical IT working environments, including an illuminating Japanese case and the working environment of IT professionals in Japan. Section 4.4 explores organisational and social measures required for the construction of an appropriately professional working environment.

4.2 Importance of a Professional Outlook to IT Professionals

4.2.1 The Notion of Profession and IT Professionals

The word profession has various meanings, from broad to narrow. In academic fields, this word is used in a restricted sense, which may be summarised into the following criteria or characteristics (Flexner 1910, 1915; Johnson 2001; Kizza 2003; Yamada 1998):

- A highly specialised body of knowledge and technique: members of a profession have an advanced, systematic, and exclusive body of knowledge, as well as techniques acquired through long-term education and training; furthermore, they continue to derive their raw material from science and learning.
- Autonomy with responsibility: professionals apply knowledge and techniques to problems freely and autonomously, assuming substantial personal responsibility; they are governed by a developed sense of personal discretion.
- Self-organisation: the social and personal lives of professionals tend to be organised around a professional nucleus; professional associations or groups are organised in order to set definite and practical ends, to set standards for practice, and to control the qualifications related to the profession and its membership based on its ends and responsibility.

- Public service: the professions have assumed an increasingly altruistic motivation, taking on the aim of working in the public service or to fulfil the profession's social function.

While the term IT professional or computer professional is regularly used in everyday life, many researchers do not consider work related to IT and information systems 'professional' in this restricted sense. For example, Hodges (2001) states that work done on computers by specialists and users is so diverse that standards of excellence, notions of success, and internal rewards are not common within the community. This means that there can be no community of values, or agreement on standards of behaviour, which constitutes the foundation for a sense of professional obligation.

Linderman and Schiano (2001) have also claimed that the field of IT cannot be a profession because it does not meet some of the defining conditions for a profession, such as certification standards, agreement on educational requirements, and meaningful or enforceable sanctions for unprofessional behaviour. Chief information officers (CIOs) are often not promoted from a group of colleagues, and may not have appropriate qualifications; in addition, priorities are often based on industrial and market interests. Consequently, those who consider themselves to be IT professionals may encounter identity problems and a power vacuum, which may lead in turn to a vacuum when it comes to social responsibility.

4.2.2 Professional Attitudes and the Stature of IT Professionals

Even though work related to IT and information systems may not constitute a profession in the traditional sense, it is not fair to say that it would be ineffectual or misleading to apply professional ethics to this evolving field. Instead, with a view to creating and preserving a safe and reliable information society and e-business environment which is the very basis of quality IT-enabled services, it is far more constructive to use knowledge yielded from the field of professional ethics to examine how the IT field can fulfil its social functions and responsibilities, and what kinds of behaviour are desirable for those IT professionals who develop and maintain IT and IT-based information systems.

For example, we could create an imaginary IT professional who behaves according to some socially accepted code of IT professional conduct as an ideal; then we could apply this behaviour to a real-world IT professional, situated within a specific context. The goal is for real-world IT professionals to develop a 'professional outlook' that underlies their code of professional conduct, because no code can be exhaustive or guarantee appropriately professional decisions, and rapid IT development could continually create novel social and ethical problems.

The development of a professional outlook and a sense of professional ethics should promote the following elements:

- Altruism: those who develop a professional outlook should recognise that their work is primarily a form of public service, and that public interest should guide their judgment and decision-making.
- Intellectual modesty: they should recognise that the quality of their work depends upon their knowledge and understand, therefore, that cognitive limitations and obsolescence of knowledge can reduce the quality of their work. This recognition leads to a respect for others and motivates continuous learning.
- Integrity: they should accept full responsibility for their work and remain honest with themselves and with others.

Flexner's words are particularly appropriate in the field of IT:

> But, after all, what matters most is professional spirit. All activities may be prosecuted in the genuine professional spirit. In so far as accepted professions are prosecuted at a mercenary or selfish level, law and medicine are ethically no better than trades. In so far as trades are honestly carried on, they tend to rise toward the professional level. [...] In the long run, the first, main and indispensable criterion of a profession will be the possession of a professional spirit, [...] (Flexner 1910)

4.3 IT Professionals in Business Organisations

4.3.1 Constraints on IT Professionals in Workplaces

IT professionals now have significant social responsibility; this will never diminish because of their intentional and/or unintentional power (Huff 2004) over a wide range of people and groups. The development and deployment of IT and information systems has transformed society irreversibly (Murata 2001), and the decision-making and value judgments that IT professionals embed in the IT and information systems they develop constitute an invisible factor in this transformation.

It is impossible to ignore the fact that the majority of IT professionals work for business organisations; they work in the context of a market economy and a business structure. Sometimes, they may be coerced into following 'logic of business' that causes them to lose touch with the public interest.

Usually, IT professionals in workplaces are under two types of constraint: contractual and intellectual. The latter involves human factors such as limits to cognition and knowledge; this is inevitable for both IT professional individuals and groups. The former relates to the multiple roles played by an IT professional in the workplace; he/she is required to follow working regulations as an employee, to abide by a code of professional conduct as a professional, to meet due dates on a budget as a contractor, to support a household as a member of a family, and so on.

These constraints often prevent IT professionals from developing their sense of professional ethics and outlook, thereby constraining their sense of responsibility and accountability. For instance, problems in software codes caused by many hands and diehard bugs, which are typical barriers to IT professional accountability (Nissenbaum 1994), could be avoided if IT professionals had no limits in cognition and knowledge or could spend unlimited time and money. However, because IT professionals play many roles, conflicts can arise between responsibilities to different stakeholders, making it difficult to maintain a sense of professional responsibility (Johnson 2001, pp. 74–76); in addition, time constraints may give IT professionals an incentive to disregard democratic values and to make a decision selfishly, or one based on the economic and political power of stakeholders.

4.3.2   Working Environment of IT Professionals

Complicated situations related to responsibility and accountability never lighten the ethical burden of IT professionals. However, a highly stressful and physically demanding working environment can disrupt the professional outlook of IT professionals, causing them to have an irresponsible or an apathetic attitude.

IT professionals in business organisations do not operate in a vacuum, and are not necessarily independent and unchallenged. Often they work in complicated situations with conflicting responsibilities, and it can be difficult for them to appropriately prioritise their professional responsibilities. IT-based information systems are often developed within tight schedules and tight budgets, with minimum number of personnel to meet a deadline. These factors may prevent developers from addressing technological, social and ethical issues relating to their information systems, which would seriously deteriorate the quality of the systems and services the systems enable.

IT workers also tend to have a precarious position within business organisations; in modern global capitalism, where investors are relatively powerful in relation to business organisations as compared to other stakeholders, many business organisations now adopt personnel policies centred on improving labour productivity and reducing personnel costs, forcing longer working hours and less rewards on all office workers. IT has been integral to 'reengineering' business processes, the result of which has been a reduction in redundant personnel since the early 1990s. Today it is IT professionals who are threatened by cost-cutting employment policies that recommend replacing full-time employees with contract workers or temporary staff, and experienced IT professionals with fresh university graduates. Rapid IT development, which has been described in 'dog years', may provide human resource managers with an excuse for the dismissal of experienced IT professionals; only those who have knowledge about state-of-the-art IT are considered indispensable. Because IT professionals produce information, which can be immediately transferred anywhere via the global net, offshore employment or global outsourcing of labour can also threaten the status of IT professionals in developed countries. Consequently, IT professionals are often employed in a highly stressful and physically demanding business environment, and it is likely that many are more concerned with their job status, personal obligations, and retirement than with social responsibility.

Such a difficult work environment may affect the judgment of IT professionals, and may cause irresponsible and apathetic attitudes, even in those who consciously intend to abide by codes of professional conduct; it may also undermine the professional spirit underlying these codes. Therefore, the construction of an appropriately professional environment, designed to encourage the development of professional attitudes in IT professionals, is crucial for maintaining high quality of information systems they develop. IT professionals employed in a positive business environment would be more motivated to uphold their social responsibilities.

4.3.3 Working Environment of IT Professionals in Japan

4.3.3.1 Imamichi's Eco-Ethics and IT Professionals

Imamichi (1989, 1990, 1998) described the necessity of developing appropriate ethics for the current technological society in his insightful studies of eco-ethics. The practical syllogism described in Aristotle's Nichomachean Ethics must be reconsidered in the modern eco-environment or human habitat, which is composed not only of nature, but of "technological conjunction". The classical form of practical syllogism is as follows.

Major premise: A is desirable.
Minor premise: p, q, r and so on realise A.
Conclusion: For some reason, I choose p as the means to achieve A.

Here, the ideal goal is obvious and the minor premise is the horizon of the freedom of choice, the object of which is a means to achieve the goal. This form of practical syllogism remains valid in our individual decisions even today. Due to the rapid progress of technology and the advent of the technological society, however, means are now more important than goals, and the logical structure of the practical syllogism has been reversed.

Major premise: We have means or power P.
Minor premise: P can realise goals a, b, c and so on.
Conclusion: We choose a as the goal of P for some reason.

Obvious powerful means exist, including nuclear, electric and electronic technology. Goals attainable using these means are considered analytically, and hence the means control the goals. The means are so powerful that a selection of a particular goal may have considerable influence on society. However, these sorts of means are usually controlled by groups or organisations, not by individuals; the subject in the modern form of practical syllogism described above is "we". This tends to result in a confusion of responsibilities for goal setting. The problem here is the *nosism* of organisations, not egoism.

These Imamichi's arguments suggest that the development of professionalism in IT professionals is vital to overcoming nosism in organisations in which IT is developed and/or used as well as to eliminating the ambiguity over responsibilities for IT development and/or use. Business organisations are usually thought to make decisions based on their productivity and economic efficiency, even though business ethics and corporate social responsibility have recently become a popular topic of conversation. In Japan, government organisations have tended not to hesitate to use technology to ensure social order and security, especially since the September 11 attacks. Under these circumstances, preserving human rights and ensuring social responsibility with respect to IT development and use have often been understated.

In addition, Japan's group-oriented culture, in which one's self-actualisation is often based on the perception of relationships with members of one's primary group (Nakane 1978), may encourage group nosism and a confusion of responsibilities for group behaviour. Developing a professional outlook and maintaining personal independence is important for those engaged in the development and use of IT to ensure socially responsible development and use of IT in Japan. However, obstacles exist for this in Japanese society as well as in the Japanese IT industry.

4.3.3.2   *Senmonka* and Professional

In Japan, there are 13 kinds of national accredited certificates relating to IT (10 for IT professionals, 1 for system auditors, and 2 for end-users). One objective of these certificates is to establish the social status of IT engineers. However, few Japanese firms require certification, even IT related firms such as hardware manufacturers, software houses, and IT consulting firms. Work experience in system development or maintenance is considered more important, and is used to evaluate the abilities of IT professionals.

Many IT employees in Japanese firms, especially firms related to IT, are called system engineers (SEs). This job title covers employees who engage in information system analysis, design and development, and project management. They are also often responsible for pre-sale and post-sale technical consulting. SEs do not have high status in Japanese firms or society. On the contrary, they are often considered 'disposable' personnel because their job is so physically demanding; in addition, they are usually required to finish their work by a scheduled completion date, which is usually set very tightly.

The Japanese word *Senmonka* is considered to correspond to the English word 'professional'. However, *Senmonka* does not exactly correspond to the term 'professional' in the narrow sense described in Sect. 4.2.1, and Japanese in general do not recognise the difference between a professional and an artisan, nor between a profession and a trade. This makes it difficult for ordinary Japanese to associate a professional with social interest.

The following case is based on a real situation. It represents how IT professionals are likely to act in an ethically questionable way given certain organisational and social circumstances:

DAMEMOTO, a large Japanese automobile manufacturer, decided to replace its outdated mainframe-based information system with a state-of-the-art C/S system. DAMEMOTO's CIO had worked in production management at DAMEMOTO's main factory for over 30 years and had been promoted to his current position 6 months ago. This project was a good opportunity for him to show his competence as CIO, so he was determined to construct a flawless information system.

A joint venture was organised to develop DAMEMOTO's new system; several experienced system engineers from the four software houses involved in the joint venture were ordered to join the project team. A clause in the contract stated that they must provide the CIO with a semi-monthly report of 'the bug control curve' to help him follow the project's progress. The bug control curve was an application of a quality control (QC) measure which was commonly used in Japanese automobile factories; the CIO proposed it, based on his experience in production management.

The CIO told the project team that he expected the number of bugs in the system programs to approximate a logistic curve. That is, if the project were well managed, the number of bugs detected would diminish as the project progressed, and would asymptotically become zero by the end of the project. Conversely, if the number did not diminish, the CIO would consider this to be evidence of the project management's failure.

However, it is impossible to control the number of bugs detected during this type of project. The project team members decided to insert bugs intentionally in the programs they coded and to 'control' the number of bugs detected against their will. As intended, the shape of their bug control curve was nearly a logistic curve.

4.3.3.3 Requirements for IT Professionals in Japan

A top priority of Japanese national IT strategic policies since 2001 has been the development of IT professionals such as software engineers (including IT architects and embedded-software specialists), digital content creators, project managers, information security specialists, IT co-ordinators and IT researchers. This emphasis is based on a recognition that the current shortage of such human resources will continue, even though they are critical to maintaining and improving Japan's international competitiveness. Recruitment of qualified foreign IT professionals is recommended for the same reason. The Japanese business community has supported the government's view on the subject. For example, the Nippon Keidanren (Japan Business Federation) issued a policy proposal in December 2007 that recommended setting up a national centre for developing advanced IT human resources (Japan Business Federation 2007).

The "Skill Standards for IT Professionals" published by the IT Skill Standards Centre, a division of the Information Technology Promotion Agency that administers certificate exams for IT professionals, defines a professional as follows.

A professional is a person who successfully achieves practical business results and contributes to the growth of the industry. A professional should:

– Achieve commitments to his/her customers and company,
– Train and develop subordinates to learn from his/her experience and knowledge,
– Continuously strive to improve his/her own business capability, and
– Be socially responsible and committed to ethical professional standards.

A professional achieves business goals and fulfils customers' requirements through a combination of appropriate skills. Advanced skills mean great value for customers, project members, partners and the company. A professional requires not only high technological skills, but also a high level of personal skills such as communication, negotiation and leadership, as well as business-related skills to deliver on this commitment to customers and the company. Moreover, a professional must to contribute to the training and development of subordinates by serving as a mentor or coach (IT Skill Standards Centre 2008).

This definition emphasises the skills a professional should have, cultivate and share. In fact, the intentional as well as extensional meaning of being "social responsible" and "committed to ethical professional standards" is not described in the report or in the relevant governmental documents, whereas the skills necessary for IT professionals are listed. In addition, well established codes of conduct for IT professionals such as the "Software Engineering Code of Ethics and Professional Practice" (Gotterbarn et al. 1999) are not well known among Japanese IT professionals who culturally tend to consider any such written code to be *Tatemae* (only for the sake of courtesy or as a formal sign of respect). Many people consider IT professional development in Japan to be limited to the maintenance and increase of a high level of technical skills. What professional development really means in a Japanese context is for professionals to learn to accept full social responsibility for their work, with a duty to their employers and customers.

4.3.3.4 Requirement to Nurture Professional IT Human Resources

Providing IT-based solutions to end-user companies has become an important business for the Japanese IT industry. IT is an enabler of these end-user companies' successful business processes and they demand effective IT-based solutions. The quality of IT-based solutions provided to end-user companies is a critical factor in the quality of products and services they supply. The quality of products and services provided to their customers determines the quality of the customers' business activities or the quality of life of the individual customers. Therefore, ensuring the quality of IT-based solutions is important for both the IT industry and the end-user companies in maintaining their competitiveness as well as for an increasingly affluent society.

The provision of quality IT human resources is undoubtedly a key factor in ensuring the success of effective IT-based solutions. Nurturing highly qualified, skilled and knowledgeable IT human resources and developing their professional outlook is an urgent social issue that should be addressed through a collaborative relationship among industry, government and academia. However, the current estimated shortfall of 150,000 IT workers has been alleged as being responsible for the deterioration of software quality and frequent information system failures. In addition, the IT profession has already become unpopular among young people in Japan. Results of a survey of 10,299 Japanese university students seeking employment showed that just 4.2 % of them wanted to work in IT departments (Mainichi Communications 2008). Even the number of people wanting to pursue IT-related university degrees is decreasing. This situation is due to the poor working conditions in the Japanese IT industry.

4.3.3.5   3K or 7K Workplace

Many Japanese companies have tried to improve their cost structure since the burst of the economic bubble in the early 1990s. In-house IT departments, which had often been considered as cost centres, became targets of restructuring and many of them were spun off into separate companies. The outsourcing of information system development, operation and maintenance, even offshore, is already commonplace. Japanese IT companies now compete with domestic as well as overseas IT companies, and keenly feel the need to reduce costs.

The pressure to reduce the personnel cost has made the working environment worse in terms of pay and working hours. The number of full-time workers is decreasing even as the number of part-time workers and contractors is increasing. The working environment of IT professionals in Japan is no exception, and conditions are often described as "3K", from the three Japanese phrases *Kitsui* (physically hard), *Kyuryo ga yasui* (low pay) and *Kaere-nai* (cannot go home). This has recently expanded to "7K" with the additional four Ks being *Kyuka ga tore-nai* (cannot take a holiday), *Kisoku ga kibishii* (stringent working regulations), *Kesho ga nora-nai* (have rough skin due to irregular hours and lack of sleep) and *Kekkon deki-nai* (not marriable).

Japanese IT professionals do receive above-average salaries. However, aside from Kyuryo ga yasui, the rest of the 3K and 7K clearly apply to the working conditions in the Japanese IT industry. That IT professionals feel their pay is low is understandable, considering the heavy burden of their work. Moreover, many of them must work overtime without extra pay. A survey of 2,214 IT professionals indicated that one-quarter of them felt that their jobs were not worthwhile, and 72.6 % recognised their profession was unpopular among young people (Nikkei Computer 2006).

Development of professionalism is difficult because Japanese IT professionals lack pride in their profession. The poor labour conditions in the field are so notorious that young Japanese people tend to shy away from becoming IT professionals. Thus, the Japanese IT industry finds it difficult to hire good workers. The result is a small number of capable IT professionals who are always forced to shoulder a heavy workload.

4.3.3.6 Business Practices in the Japanese IT Industry

The Person-Month Myth

Regarding the business practices of the Japanese IT industry, several obstacles exist to the development of IT professionalism. The "person-month" is still used as the usual measure of development cost for information systems. This was originally used in the manufacturing industry and is not necessarily a good fit for estimating the product cost of human intellectual activities such as software. As described by Brooks (1995), the concept of the person-month assumes that an IT professional is productive all the time. It also presumes that one IT professional can be substituted with another, ignoring any difference in skills and knowledge between them. These assumptions are based on the misunderstanding of intellectual work and workers.

This unrealistic method of cost estimation often results in unreasonable information system implementation deadlines, and turns gold-collar workers into blue-collar or entry-level workers. Because punctuality is the soul of the Japanese business culture, the frontline members of system development teams are often forced to work without relief to meet deadlines while sacrificing their job quality. This requires physical capacity and mental toughness, not creativity or original thought.

In addition, the terms and conditions of a written contract for the development and implementation of information systems in Japan are sometimes considered to be *Tatemae*, a simple formality with no real substance. In general, a written contract is prone to be considered far less important than the mutual faith and trust of all parties to the contract (Nakane 1972), and a conflict between them, if it occurs, tends to be resolved without resorting to the provisions of the contract (Uchida 2000). The Japanese business culture of "the customers are our gods" or "the customers are always right" encourages this tendency, and scope creep without adjustment of deadlines is not unusual.

Under these circumstances, a more competent IT professional in a system development team must work harder and longer to meet an immovable deadline, which eventually tends to reduce the quality of the work performed. Hence, even though the quality of work should be dependent on the worker's personal skills, taking full responsibility for personal performance becomes very difficult.

False Contracting and Multi-tier Subcontracts

Recently, false contracting has become a serious social issue in Japan in which contract-based workers are forced to work for their client companies as if they were normal employees. This is common in the Japanese IT industry to reduce personnel costs, even though it is illegal. Moreover, many Japanese IT professionals do not understand that false contracting is illegal.

Multi-tier subcontracts are also commonplace. One project manager told the author that he had worked on a project having 16 levels of subcontracts. The situation in which an engineer working for a third-tier contractor, for example, is sent off to work for the prime contractor or end-user company due to a shortage of skilled engineers, is quite common, but also illegal. These business practices make project management difficult and tend to turn IT professionals into programming machines.

4.4 Construction of an Appropriately Professional Working Environment

4.4.1 Organisational Measures to Construct an Appropriately Professional Working Environment

In order to create and preserve high quality of IT-enabled services, it is necessary not only to institute a code of professional ethics for IT professionals, but also to take organisational measures to relieve the pressure and stress that can induce unprofessional behaviour. Managers in business organisations need to understand that an appropriately professional working environment, in which IT professionals are prompted to maintain a professional attitude and to behave in an ethical manner, has practical or strategic value as well as ethical value.

IT professionals operating in an appropriately professional working environment could contribute to improved business performance by means of:

- Their professional integrity, which could improve the quality and reliability of the goods and services they produce, thereby boosting the organisation's trustworthiness and reputation, and lowering the costs of human resource management (HRM);
- Their concern for public interest, which could mean caring about a wide range of stakeholders and, therefore, obviating risks to the organisation's trust and reputation; and
- Their high motivation for learning, which could facilitate maintenance of a high standard of knowledge and lessen HRM costs such as those incurred for in-house education.

In the e-business environment, an organisation's trustworthiness and reputation are key, but intangible, assets in the construction of relationships with business partners and customers (Murata 2003). Even in the short term, an appropriately professional working environment can be quite effective in enhancing an organisation's business performance. In the long term, a business organisation with an appropriately professional working environment could enjoy favourable relationships with its business partners, customers, and its skilled and loyal employees, helping the organisation to remain competitive.

The following measures could be applied to take pressure off IT professionals and to make their working environment less stressful: guaranteeing IT professionals enough income for their living expenses, including such expenses after retirement; guarantee of status; recruitment of a sufficient number of employees; clear presentation of the typical career path; intrinsic motivation through challenging tasks; manifestation of the business organisation's values; and provision of mental health care.

In order to construct such a working environment, whereby IT professionals would be supported in conducting their duties with professionalism, both organisational business measures and social measures are required.

4.4.2 Social Measures to Support an Appropriately Professional Working Environment

An appropriately professional working environment for IT professionals is beneficial to society; poor working conditions tend to cause fatigue and apathy in them, which can lead to insensitivity to technological, social and ethical issues, ensuing, ultimately, in failures within society. Conversely, because IT development and use now underlie such a wide range of economic activities, the production of reliable, quality goods and services by IT professionals with a developed sense of professionalism can contribute to the activation of economic activities. In this way, support for the construction and maintenance of an appropriately professional working environment is socially meaningful.

The following measures may be effective in supporting such a working environment for IT professionals: establishing official certification for IT professionals; forming a trade union of IT professionals across businesses; creating legislation to protect whistle blowers; and setting up social safety nets for IT professionals.

Improvement of the IT working environment is a major requirement for improving IT professionalism in Japan, and this is an urgent social issue. One way to do this would be to create an independent industry-wide organisation to support the improvement of the social status and the working environment of IT professionals. The Programmers Guild (http://www.programmersguild.org/) would be a good example of this. However, to ensure the effectiveness of the activities this organisation undertakes, Japanese IT professionals who join the organisation must recognise it as being their primary group. The tradition of co-operative in-house unions in Japan would be an obstacle to the effectiveness of such an organisation.

Developing self-motivation among Japanese IT professionals is a prerequisite for improving the IT working environment. However, almost all IT professionals are salaried employees, and the amount of their pay is not based on their performance. Any pay differences among members of a system development team are intentionally small; significant differences would degrade the team effectiveness due to the Japanese egalitarian culture. Japanese IT professionals are usually motivated to take responsibility for their teams and companies, not for their performance, meaning that Japanese business practices would have to change.

On the other hand, one of the most serious problems with respect to business practices in the Japanese IT industry is that written contracts often do not have practical force. This is true not only for development contracts between companies, but also for employment contracts between IT professionals and the companies hiring them; moreover, the contracts tend to be downplayed. In the industry, the contents of implicit contracts are often stretched, with the more powerful parties to the contracts holding the advantage.

Examining how written and implicit contracts should function in the Japanese IT industry is an urgent issue to improve business practices and thus to establish a reasonable level of IT professionalism. Everyone in the industry must understand the importance of the contracting process (Uchida 2000) and develop an appropriate respect for contracts.

4.5 Conclusion

In an information society, IT professionals in business organisations have power over the general public's quality of life. Thus, IT professionals are responsible to the public and need to develop a professional outlook and ethical attitude in order to attain a high level of quality of IT-enabled services and consequently to create and preserve a safe and reliable information society.

However, individual IT professionals are not necessarily independent and unchallenged individuals; their behaviour can be affected by stress and pressure experienced in the workplace. Accordingly, organisational as well as social measures are needed to construct an appropriately professional working environment in which IT professionals are supported in maintaining their professional ethics and outlook.

An appropriately professional working environment has practical value as well as ethical value for business organisations and IT-enabled services they provide, and is beneficial to society. Such a working environment could constitute the basis for professionalism in IT professionals, which, in turn, would provide the basis for a safe and reliable society. The efficacy of codes of professional conduct can only be ensured through the construction of an appropriately professional working environment.

Acknowledgments This study is supported by the MEXT (Ministry of Education, Culture, Sports, Science and Technology, Japan) Programme for Strategic Research Bases at Private Universities (2012–16) project "Organisational Information Ethics" S1291006 and the MEXT Grant-in-Aid for Scientific Research (C) 22530378.

References

Brooks FP (1995) The mythical man-month: essays on software engineering. Addison-Wesley, Reading

Flexner A (1910) Medical education in the United States and Canada: a report to the Carnegie Foundation for the Advancement of Teaching. Bulletin Number Four (The Flexner Report), The Carnegie Foundation for the Advancement of Teaching, available online at http://www.carnegiefoundation.org/publications/medical-education-united-states-and-canada-bulletin-number-four-flexner-report-0. Accessed 25 May 2012

Flexner A (1915) Is social work a profession? In: Proceedings of the national conference of charities and corrections, pp 576–590, available online at http://ia600406.us.archive.org/10/items/cu31924014006617/cu31924014006617.pdf

Gotterbarn D, Miller K, Rogerson S (1999) Software engineering code of ethics: approved! (Software engineering code of ethics and professional practice, Version 5.2). Commun ACM 42(10):102–107

Hodges MP (2001) Does professional ethics include computer professionals? Two models for understanding. In: Hester DM, Ford PJ (eds) Computers and ethics in the cyberspace. Prentice-Hall, Upper Saddle River, pp 195–203

Huff C (2004) Unintentional power in the design of computing systems. In: Bynum TW, Rogerson S (eds) Computer ethics and professional responsibility. Blackwell, Malden, pp 98–106

Imamichi T (1989) The concept of an eco-ethics and the development of moral thought. In: Tnag Y, Lie Z, McLean GF (eds) Man and nature: the Chinese tradition and the future, the council for research in values and philosophy. http://www.crvp.org/book/Series03/III-1/chapter_xv.htm. Accessed 2 June 2008

Imamichi T (1990) Eco-ethica. Kodansha, Tokyo (in Japanese)

Imamichi T (1998) Technology and collective identity: issues of an eco-ethica. In: Imamichi T, Wang M, Liu F (eds) The humanization of technology and Chinese cultures. The council for research in values and philosophy. http://www.crvp.org/book/Series03/III-11/chapter_i.htm. Accessed 6 June 2008

IT Skill Standards Centre (2008) Skill standards for IT professionals, Version 3, Part 1: Overview. http://www.ipa.go.jp/english/humandev/second.html#v3. Accessed 16 June 2008 (in Japanese)

Japan Business Federation (2007) Towards acceleration of advanced ICT human resources: proposal for setting up the National Centre. http://www.keidanren.or.jp/japanese/policy/2007/106/honbun.html. Accessed 12 Feb 2008 (in Japanese)

Johnson DG (2001) Computer ethics, 3rd edn. Prentice-Hall, Upper Saddle River

Kizza JM (2003) Ethical and social issues in the information age, 2nd edn. Springer, New York

Linderman JL, Schiano WT (2001) Information ethics in a responsibility vacuum. Database Adv Inform Syst 32(1):70–74

Mainichi Communications (2008) Survey on University students' job seeking behaviour. http://job.mynavi.jp/conts/saponet/enq_gakusei/ishiki/ishiki08/2008ishiki.pdf. Accessed 26 June 2008 (in Japanese)

Murata K (2001) Social and ethical aspects of IT. Off Automat 22(3):30–35 (in Japanese)

Murata K (2003) Trust and reputation as corporate assets and information ethics. In: Tohyama A (ed) Post-IT strategy: from e-business to business. Nikkagiren, Tokyo, pp 163–185 (in Japanese)

Nakane C (1972) Conditions for adaptation. Kodansha, Tokyo (in Japanese)

Nakane C (1978) Dynamics of the vertical society. Kodansha, Tokyo (in Japanese)

Nikkei Computer (2006) Questionnaire survey on ICT professionals' work. Nikkei Comput (668):38–53

Nissenbaum H (1994) Computing and accountability. Commun ACM 37(1):73–80

Uchida T (2000) The age of contract. Iwanami Shoten, Tokyo (in Japanese)

Yamada R (1998) Professional school: professional education in the USA. Tamagawa University Press, Tokyo (in Japanese)

Chapter 5
A Community Based Trust Establishing Mechanism for a Social Web Service

Shigeichiro Yamasaki

5.1 Introduction

Encountering new people is indispensable for succeeding in business and enriching the life of a person. Communities and companies that fail to take on new challenges and invite new participants have difficulty surviving. The primary purpose of a trust infrastructure is to reduce the risk caused by new encounters and to give us the freedom to challenge new businesses and make new friends. IT enabled services subject to this chapter are social Web services. Social Web services offer the possibility to amplify a person's ability to meet new business and new friends. In order to make this power actually available, it is necessary to provide an infrastructure of trust for social web services.

5.1.1 The Infrastructure of Trust in the Age of Social Web Service

Before the social web services become popular, primary information source in the Internet was the web pages. Search engine services such as Yahoo!, Google has been succeeded by providing the effective means to search valuable web pages.

Social web services has become popular from 2003. Those services had obtained a huge population of users in a short period of time. Facebook had gotten more than 800 million users at June 2011. Enormous number of people have come to disseminate information over the social Web service. Recent years, the primary information source in the Internet has become the people on the social web service from a web page.

S. Yamasaki (✉)
Department of Information and Computer Sciences, Kinki Uiversity, Kyanamori 11-6,
Fukuoka, Japan
e-mail: yamasaki@fuk.kindai.ac.jp

S. Uesugi (ed.), *IT Enabled Services*,
DOI 10.1007/978-3-7091-1425-4_5, © Springer-Verlag Wien 2013

Usefulness of the Web page is based on the described content of it. However, the usefulness of a person on the social web service as an information source is the expectation for the information which the person will inform in the future.

Intuitive definition of "trust" in this chapter is the expectations for reliability and usefulness of the information a person will disseminate. Our motivation is to present a platform to evaluate the trust of the individuals and companies as an information source over the social Web services.

5.2 The Concept of Trust

There are various definitions for the concept of trust. First, we will try to organize the concept of trust.

5.2.1 Trust for Governance and Trust for the Market

There are two types of trust based on the purpose of it: trust for the market and trust for governance. The purpose of the trust for governance is that a certain range of members are allowed safe interaction with each other under the guarantee of an authority. The purpose of the trust for the markets is to minimize the risks for companies that do business in the open market where anyone can participate.

Trust for governance limits the number of members to be included. On the other hand, the trust of the market does not exclude any person. The trust of these two, in fact, are neither independent or mutually exclusive. In order to make its foundation firm, trust for the market uses the trust for governance. However, we should not confuse these different meanings of "trust".

5.2.2 Definition of Trust in Social Psychology

Social psychology also has a lot of proposals for the definition of the concept of trust. Yamagishi defined trust as "expectations for the order of the social morality" (Yamagishi 1998). Furthermore, Yamagishi gave a more detail definition that includes: (1) expectations for the party's ability to accomplish it, and (2) expectations for the willingness to perform the responsibility of the party. Yamagishi also pointed out that the (3) meta-communications needed for the trust to state to the party, "I believe you", is an important element for establishing trust (Yamagishi and Yoshikai 2009). In the area of finance, an example of trust in the ability of a financial person is his/her income, an example of the trust in their motivation is the mortgage, and an example of the trust in the meta-communication of a financial person is the financial contract.

We adopt these three definitions of trust for our discussion.

5.2.3   Fostering Trust Through the Reputation of an Open Community

The appropriate style of the foundation of trust varies depending on the era, the size of a population, and the social system of a nation.

A nation has the ability and will to bear the responsibility for the governance of trust. In the first half of the twentieth century, most of the nations of the world worked well as a foundation of trust for the governance. However, in the latter half of twentieth century, many nations, including Japan, began to shrink and privatize the functions of it. For example, in Japan from 1995 through 2009, the number of civil servants had decreased by 894,000. By reducing the function of the nation, the ability to accomplish trust through the nation has also been reduced.

Contrastingly, in recent years, giant enterprises of social web services like Google, Facebook, Twitter, etc. have appeared. These giant enterprises have the ability and will to bear a part of the responsibility of trust, to get new users, and to have the users continue using their services. However, the trust that they provide is minimum.

On the other hand, on social Web services, the users and their community have a more powerful ability and strong will to bear the trust because they seriously need it. Therefore, it is natural that we expect the foundation of fostering trust through the reputation of an open community over the social web.

5.3   The Concept of Persona

Next, we describe the concept of a "persona," which is another central concept in this chapter. A persona is intended to symbolize the different side of the social role of a person. A "pseudonym" is an identity of a person which can guarantee their attributes while hiding their real name. A persona can also be used as a pseudonym if it is used with an authentication infrastructure and with a well designed attribute authorization system.

5.3.1   Persona and Privacy

A persona is useful for controlling the range of personal information. We are able to bring together the attributes and behaviors of a person with an independent persona which is a representation of a person's role in the workplace, in the family and in their relationships with friends. By selecting the appropriate persona for the scene we can control private information.

Some kinds of persona may be used for the exposure of illegal acts. Such kinds of persona will be created and destroyed temporally by their owner.

5.3.2 Accumulating the Reputation of a Persona

A persona can also be an means for accumulating one's reputations. Notice that, because everyone who owns a persona may destroy and re-create it freely, the negative reputation of the persona will not work well. However, there is an incentive to keep a persona that has a positive reputation that can accumulate ratings. A company in the social Web services is one persona. The persona of a company can also obtain trust by accumulating positive a reputation.

The important thing to create consistency for a persona is that the value of the reputation which has been accumulated by that persona reaches a level that is too high to throw away. Long term time lines of a persona over the social web become evidence for proving that the persona has exited in an active state (Fig. 5.1).

5.3.3 Persona and Cross-Web Authentication and Authorization

A trustable IT enabled service should be equipped with the functionality (1) to confirm persons who interacts in the real world (2) to confirm the consistency of persons who have been interacted with and (3) to confirm the privileges of the persons who interact with them. User-centric authentications and authorizations such as OpenID or OAuth are made use of through cross-website safe mash-up for personal data exchange. Additionally, OpenID connect has also been specified for integrated cross-web site authentication and authorization (Passant et al. 2009). We can understand a persona as a symbol for (1), (2) and (3) when all the functionality of user-centric cross web-site authentication and authorization mechanisms works (Sakimura et al. 2011).

5.3.4 Open Persona

The essence of privacy and copyright is the right to control information which is owned by a person. To guarantee certain rights to control copyrighted literature and private information of individuals is important. However, too strong rights about those cause serious problems rather than convenience. The works of open source software have succeeded by deliberately weakening the rights of the owner and liberalizing access to the works. In the same way, we suppose that a person classifies and separates personal information and organizes them with personas. Also, we assume that some of the personas intentionally weaken the rights of privacy and whose information is accessed freely. We call this "open persona."

Fig. 5.1 Negative rating and positive rating of persona. A persona with a negative rating can be destroyed and reproduced. A persona with a positive rating will be maintained perpetually

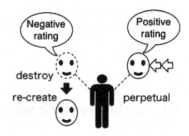

By taking advantage of the open persona, we can control the scope to disclose personal information. For example, when you want to receive the recommendation of the books only a certain field in the book purchase site, you can create an open persona only for this purpose. An open persona is better in the following points than an automatic recommendation system.

1. Range of personal information disclosed by an open persona can be controlled by the will of the person.
2. By continuing to use the persona of a particular application, the accuracy of The recommendations for the person will be improved.
3. To configure a community of highly similar personas is special fields, the accuracy of recommendation systems can be improved.

5.3.5   Disclosure and Trust Game

Trust is not required in a closed society. Trust is needed in order to obtain a new business opportunity and meeting new people in an open society. When meeting a new person, disclosure of the party is an important factor in order to trust the party. Similarly, to get trust of the opponent, one has to disclose the private information of oneself.

On the other hand, the disclosure of private information involve risks. Actually, there exist many malicious social web applications which aim to exploit private information. In addition, this principal does not disclose their information, after taking the opponent's personal information, they may destroy their persona. Therefore, partners who want to form trust relationships should only be persons who disclose their information properly. This also applies to the party, all people should disclose their own information if the they want to be trusted.

Thinking in game theory, two people who want to form a trust relationship with each other would get the highest gain when disclosing the same level of personal information. As there is no deduction from the gain even if just taking the information of an open persona, the strategy that both sides use open personas is a Nash equilibrium.

5.4   Technology Related to the Reputation Based Trust

Reputation based trust is to perform the evaluation of the trust of a person based on the records of the interactions among other people.

5.4.1   Related Works of the Reputation Based Trust

Before beginning the discussion about reputation based trust, we will discuss related works.

Artz and Gil (2007) surveyed the researches about trust in computer science and the Semantic Web. According to their study, research about trust is classified into the following four categories, (1) based on the certificate issued by the trustable principal, (2) based on reputation by the community, (3) based on the calculated value based on the general model of trust, and (4) based on the study of trust of information sources. Furthermore, they classified the researches based on reputation by the community as follows, decentralization and referral trust, trust in P2P networks and grids, and trust metrics in a web of trust.

Decentralization and referral trust

For decentralization and referral trust studies of distributed management against security problems for the centralized management of reputation, in this study we include "open networks by (Beth et al. 1994) is an algorithm to evaluate the reputation of a P2P system based on the transitivity of trust.

Trust in P2P networks and grids

For trust in P2P networks and grids studies on the evaluation methods for the quality of the data without guarantee over P2P networks, the EigenTrust algorithm (Kamvar et al. 2003) is an algorithm to evaluate the reputation of the P2P system based on transitivity of trust. The reliability of each peer is computed with the PageRank algorithm (Brin and Page 1998; Langville and Meyer 2006). Damiani et al. proposed an automatic voting protocol called XRep (Damiani et al. 2002a, b) to select the best host to get a reputation for the resource from the feedback.

Trust metrics in a web of trust

> For trust metrics in a web of trust studies on trust as a distance which is calculated from the transitive relationship through the mutually related information over the web, Golbeck and Hendler proposed the TrustMail (Golbeck and Hendler 2004a, b) system.
>
> In their system, each entity fixes its reputation based on the information of other entities. They call it a 'web of trust.' They are using an ontology to represent trust and reputation. This enables calculation of the degree of trust between any two entities. They call it a "trust metric."

5.4.2 Trust Based on the Reputation Over the Social Web

5.4.2.1 Trust Relationship Based on Follow-Up Relationship of Twitter

Google's PageRank (Brin and Page 1998) is the most famous method to evaluate web pages. It is similar to the evaluation method for literature which uses the number of citations. PageRank evaluates the value of a web page from the number of hyper links. A lot of methods similar to PageRank have been proposed for the evaluation of trust of persons in a social web like EigenTrust (Kamvar et al. 2003). We have tried to evaluate the trust of a person from the number of followers on Twitter. We regard the following relationship as a recommendation of a person. We call this method "PersonaRank" (Yamasaki 2010).

The features of PersonaRank are: a person that is followed by many people will be valuable persons, and a person who is followed by a valuable person will also be a valuable person. The calculation model of PersonaRank is almost the same as Google PageRank. The outline of the calculation is as follows (Fig. 5.2).

P_i: Persona i

$r(P_i)$: The PersonaRank of P_i

F_{P_i}: The set of followers of P_i

$|P_i|$: The number of people following P_i

H: Sub − stochastic following relations matrix

G: Google matrix

a: lonely node vector

α: Scaling parameter, a scalar between 0 and 1

e: Row vector (every element is 1)

π^T: PersonaRank vector

$$\pi^{(k+1)T} = \pi^{(k)T} \mathbf{G} \tag{5.1}$$

Fig. 5.2 The features of our PersonaRank are (*1*) a person that is followed by many people will be a valuable person. (*2*) a person who is followed by a valuable person will also be a valuable person

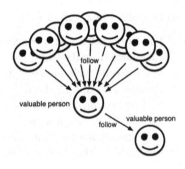

$$G = \alpha H + (\alpha a + (1 - \alpha)e)\frac{1}{ne^T} \qquad (5.2)$$

$$r(P_i) = \sum_{P_j \in F_{P_i}} \frac{r(P_j)}{|P_j|} \qquad (5.3)$$

$$H_{ij} = \frac{1}{|P_j|} \qquad (5.4)$$

5.4.2.2 A Subjective Trust Evaluation Method from the Similarity of the People

Trust is a subjective concept. Therefore, it is natural that the results of the evaluation of trust are different according to the evaluator. For example, a person who loves some special type of wine will feel they can trust another person who loves similar types of wine.

We utilize the similarity between the evaluated and the evaluator person to reinforce the evaluation of trust. We use collaborative filtering technology (Goldberg et al. 1992) to calculate the similarity of the people. Collaborative filtering technology is famously used in Amazon's recommendation system. Similarity of behavior between Twitter users are defined as follows using the Pearson correlation coefficient.

$$p(x, y) = \frac{\sum_{i=1}^{n}(x_i - \bar{x})(y - \bar{y})}{\sqrt{\sum_{i=1}^{n}(x_i - \bar{x})^2}\sqrt{\sum_{i=1}^{n}(y_i - \bar{y})^2}}$$

$$Dist(P_i, P_j) = 1 - \frac{p(P_i, P_j) + 1}{2}$$

5.5   Problems of Trust Based on Statistical Information

5.5.1   Result of the Experimentation of Trust by the Following Relationship

We collected the data of 28,830 Twitter users randomly and calculated PersonaRank. According to the score of PersonaRank, we ranked the users. The top 5 highest scoring users are listed as follows.

1. "Britney Spears"
2. "ashton kutcher"
3. "Ellen DeGeneres"
4. "Lady Gaga"
5. "Barack Obama"

Almost all persons on this list are famous people. It is natural for many people follow a famous person. However, these famous people are not always trustworthy or valuable as a information resource.

The intention to follow someone in the social web and to link some web pages are different in terms of the trust for those information sources. Most of the messages on the time line of Twitter contain chatting or jokes. However, some people appreciate famous people's chat and joke. The value of the information of the social web and web pages are quite different.

In any case, PersonaRank or other information rating methods based on the number of followers or friends can not be a suitable index for measuring the trust for the information sources over social web services.

5.5.2   Experimental Results of Trust Evaluation Based on the Similarity of Behavior

Next, we confirmed a trust evaluation method which focuses on the similarity among users. We have focused on the temporal similarity of behavior among people as an index of similarity. We have selected 578 active users from 28,830 users, who had posted more than 20,000 tweets. We had recorded for 1 week and counted the number of tweets per hour (Fig. 5.3).

After correcting that, we calculated the similarity of every pair of the people. The graph of Fig. 5.4 shows the similarity of the top 30,000 pairs. After checking the results of our experimentation, almost all the pairs of high similarity turned out to be a BOT. A BOT is a software robot, which imitates human behavior. We found that BOTs are prevalent amongst social web services. This result has serious implications because the interest of this research is concentrated on the pairs of high similarity and it is very hard to distinguish the BOTs from humans automatically.

Fig. 5.3 Frequency of tweeting as to time and persona. The columns of this table represents the persona and the row of the table represents the time from Monday to Sunday. The content of each cell represents the times of tweets in 1 hour

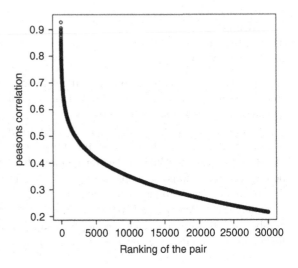

time		P1	P2	P3	P4	P5	P6
monday	0AM	2	8	7	0	2	14
monday	1AM	12	5	1	2	5	17
...							
sunday	10PM	4	4	7	9	12	22
sunday	11PM	6	7	16	11	0	12

Fig. 5.4 The relationship of the similarity and the number of pairs. This result shows that there exists small number of pairs of very high similarity between users

This problem is not restricted to the similarity of lifestyle. Activity BOTs and the technologies that imitate human conversation have become sophisticated. After examining such kinds of BOT technologies, we concluded that clever BOTs would skew the data in any statistical approach. Such kinds of problems would still be hard to avoid even when using authentication technologies because some actual person or company authenticates most of the BOTs.

5.6 Trust by Meta-Communication

The most serious shortcoming of the above approach is that we considered trust to be represented by statistical information over the social relationship. The human race, over time, has designed various social methods and social devices to establish trust. We have to reconfirm such kinds of deep knowledge about trust. A trust rating method requires more precise evidence of trust.

As we mentioned before, the definition of trust according to social psychology includes expectations for the party's ability to accomplish it, expectations for the willingness to perform the responsibility of the party, and meta-communications for trust to state the party. Among these definitions, the ability and the motivation can be captured as a statistical index, however, meta-communication is more human oriented.

The most basic meta-communication for trust is to say "I will trust you" from the person who is trusting to the person who is trusted. Such an assertion makes a commitment to the party. And the party carries the social responsibility when he or she responds by saying "All right" (Fig. 5.5).

5.6.1   Areas Suitable for Our Approach

The evaluation of the evaluator is also an important thing because without it the result of the evaluation of trust can not be validated. A responsible evaluation is a costly job for the evaluator. The most important prerequisite for its adoption is the existence of fair motivation on behalf of the evaluator. The proposed trust-rating model has limited application (Fig. 5.6).

5.6.2   Our Meta-Communication Protocol for Trust

It is important that parties who commence trusting each other confirm it explicitly. Some form of meta-communication, like saying, "I will trust you" and responding "All right, is required." In the real world, not only legal contracts, but also various kinds of social ceremonies, like a declaration of marriage, require a strong mental commitment from the parties and their explicit statement of trust. In our social model, when a subject wants to trust some object with which they share a weak tie of social relationship, explicit meta-communication is required. In our meta-communication protocol, the subject of trust sends a meta-communication message to the object of trust before they begin actual communication. We also assume that the subject is a state machine that changes its states with the events of this protocol.

5.6.3   The Mental States of the Subject and the Reputation of the Object of Trust

After meta-communication is complete, a subject of trust forms his/her mental model about trust for the object. When the state of the subject becomes "trustworthy" the

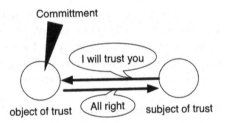

Fig. 5.5 Meta communication and the commitment of trust. When a person had stated "I will trust you" the person would receive the commitment. And the party carries the social responsibility when he or her had responded to statement "All right"

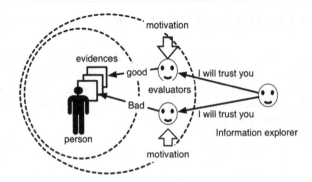

Fig. 5.6 The basic model of the meta-communication approach. This model must have the evaluation of the trust of the evaluator. The motivation to evaluate trust faithfully is also important

value of the reputation of the object of trust goes up. When the state of the subject becomes "untrustworthy" the value of the reputation of the object of trust goes down.

5.6.4 The Measurement of Trust Ratings from the Commitment Network of Trust

The algorithm of trust rating is basically the same as PersonaRank. The network structure is not the following relationship, but rather, the commitment relationship. The value of the reputation can be negative if the mental state of the subject of trust is untrustworthy.

5.7   Some Applications of the Trust by Meta-Communication

5.7.1   A Trust Rating System for the Job-Hunting System for IT-Engineers

As an example of our proposal, we constructed a job-hunting system for IT engineers as a Facebook application. IT engineers tend to publish their results of activity as web content. Examples of such web content include, source code repositories of open source software and its documents, shared slides of technical meetings, blogs of technical information and logs of technical meetings. If the personal web contents of an IT engineer are integrated to form one distinguishable name, we can use the integrated content as evidence of the ability of the IT engineer. An employer who wants to search for an IT engineer of some specialized area of technology can easily classify the candidates by the integrated web content of the candidate.

5.7.2   Mutual Evaluation Among IT Engineers and Its Motivation

Like an IT engineer in the same field, who has the ability to evaluate the skill of another, there exists a symmetrical relationship between the objective IT engineer and his/her evaluator. An evaluator also evaluated by the employees and reputation as an evaluator is accumulated as an ability of the IT engineer. This is the motivation for the evaluator to be honest (Fig. 5.7).

5.7.3   Meta-communication Using a Loyalty Program

The primary purpose of a loyalty program is to give customers an incentive to use same service. A loyalty program has another effect on customers. It is the index used to evaluate the customer as "loyal customer". A loyalty program has pseudo-bidirectionality, from customers to the service, and from the service to the customers. That is, a customer uses points as a method for settlement and the service issues points to the user. Issuing points to a user is a kind of meta-communication which means "you are our royal customer". The customer who gave points will feel a sense of superiority in being a loyal customer. It is also an another commitment. The meaning of meta-communication is not limited to trust. However, almost all meanings have similar structure. To expand the usage of loyalty programs we can expand the meaning of meta-communication.

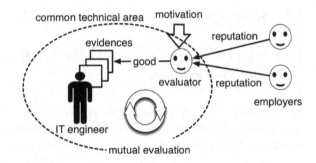

Fig. 5.7 A trust rating system for the job-hunting system for IT-engineers will work because IT-engineers have motivation to evaluate each other

5.7.4 Loyalty Program and UGC (User-Generated Content) Sharing Service

UGC (user-generated content) sharing service for images, photographs, videos, music, encyclopedia, vocaloid, etc. and its community has been developing. We can list some famous UGC sharing services like Wikipedia, YouTube, Flickr, etc. There is very high quality user-generated content. In most cases, such high quality UGCs are the result of contributions from a very large number of people who belong to the user community. An UGC community can be a issuer of a kind of loyalty program. It is not a loyalty program for the value to purchase some goods, but instead, for a value that demonstrates expectations and trust for the quality of work which the user creates.

To issue the points of a loyalty program of the community is a meta-communication, it means "We will trust you as a good creator of our community". And the user who gets the points would feel a sense of commitment and responsibility that "I am a member of the creator community".

5.8 How to Control the Phenomenon of Information Cascading

Everyone is influenced by the information that comes from the people around them. Information cascading occurs when people observe the actions of others and then make the same choices others have made. People tend to be influenced by information which has no way of being confirmed directly. Because of the occurrence and size of the phenomenon of information cascading, it is a complex system, and its prediction and control is very difficult. In the construction of a stable social infrastructure of trust based on the reputation of the people, this phenomenon of information cascading is a problem in two ways.

1. A person can become an untrusted source of information based on the influence of the people around that person.
2. The trust evaluation system becomes untrustworthy due to this phenomenon.

For the first problem, we can expect that a person who had accepted the commitment and responsibility as a evaluator would withstand the rumor. We can consider a way to control the phenomenon of information cascading to deal the second problem.

5.8.1   Merits of the Phenomenon of Information Cascading

There exists some positive side of the phenomenon of information cascading. This phenomenon is the origin of the power of information distribution. With this phenomenon, a user of social web services can get fresh and valuable information and can distribute their own ideas to the people who need them.

5.8.2   Anxiety Reduces the Thresholds of Weakly Tied People

A person without the means to confirm an object directly tends to follow another person's evaluation easily. When people are anxious, it is hard to maintain reasonable thresholds. For example, in the midst of a major disaster people feel anxious about the contamination of food, and erroneous information can be spread.

5.8.3   How to Control the Phenomenon of Information Cascading

A small amount of information cascading always occurs everywhere in small communities over a huge social web service. In a small community it is natural that some important information will spread all over the community. Sometimes small cascading triggers large scale cascading in this way. The precise conditions which cause such large cascading is not known. Because small cascading is useful, we try to control the size of cascading.

5.8.3.1   Betweeness Centrality of a Network

"Betweeness Centrality" is a popular index to represent the centrality of the network structure (Barthelemy 2004). A vertex which has a high value of betweeness centrality is a mediator of multiple communities. In the example of Fig. 5.8, v_1 seems to be central vertex because vertex v_0 has only three arcs and vertex v1 has six arcs. However, give all the pairs of the vertex in the network and

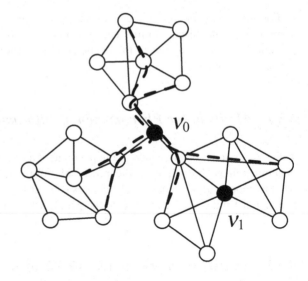

Fig. 5.8 v_0 is the betweeness center because in this network, the one whose path is crossed the most by the shortest paths of the every pair of vertex is v_0

draw the shortest path of each pair, the vertex whose path is crossed the most times is v_0. The betweeness center of the network is the vertex which is crossed the most times by the shortest paths of every pair of the network. In Fig. 5.8, v_0 is the betweeness centrality.

Betweeness centrality is calculated as follows.

$$b_i = \frac{\sum_{i_S=1; i_S \neq i}^{N} \sum_{i_T=1; i_T \neq i}^{i_S-1} \frac{g^{(i_S i_T)}}{N_{i_S i_T}}}{(N-1)(N-2)/2}$$

5.8.3.2 Division of the Network by Betweeness Centrality

Deleting the vertex of betweeness center in the network, the network becomes divided to be the set of clusters (Girvan and Newman 2002). This dividing is a coarse method of finding communities. As a method to control the size of the phenomenon of information cascading, we can use the splitting of the network by betweeness centrality (Fig. 5.9).

The most influential person is a person who has many followers. However, the most important factor of wide area information cascading is a person of the betweeness center of the social network. Therefore a person who has both property is an important person to control the information cascading.

Fig. 5.9 The vertex of betweeness *center* is a boundary of the cluster. Deleting the vertex of high betweeness centrality in the network, the network become divided to be the set of clusters

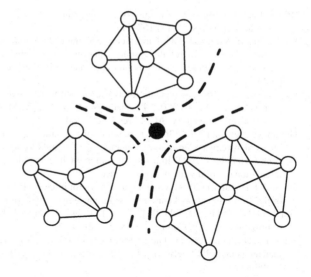

5.9   Conclusions

Social web services have the possibility to be an important social infrastructure in the future. The foundation of trust is indispensable to reduce the risk caused by new encounters and to provide the freedom to obtain new businesses and new friends for us over social web services.

The trust infrastructure for social web services based on the reputation of the people is the most natural trust foundation now. The application area of our trust mechanism is limited. Nevertheless, our proposal is part of the reconstruction of social devices for trust which have been considered by the long history of human beings. We have to learn more from this history.

Technologies used to prevent the phenomenon of information cascading are still immature. However, there is a lot promising knowledge about approaching these problems. We expect that we can progress this research and apply it to such knowledge.

References

Artz D, Gil Y (2007) A survey of trust in computer science and the Semantic Web. In: Web semantics: science, services and agents on the world wide web, vol 5, Issue 2, June 2007, pp 58–71

Barthelemy M (2004) Betweenness centrality in large complex networks. Eur Phys J B Condens Mat Complex Syst 38(2):163–168

Beth T, Borcherding M, Klein B (1994) Valuation of trust in open networks. In: Proceedings of the 3rd European symposium on research in computer security, pp 3–18

Brin S, Page L (1998) The anatomy of a large-scale hypertextual Web search engine. Comput Networks ISDN Syst 30(17):107–117

Damiani E, di Vimercati DC, Paraboschi S, Samarati P, Violante F (2002) A reputation-based approach for choosing reliable resources in peer-to- peer networks. In: CCS'02: Proceedings of the 9th ACM conference on computer and communications security. ACM Press, New York, pp 207–216

Damiani E et al (2002) A reputation-based approach for choosing reliable resources in peer-to-peer networks. In: CCS'02: Proceedings of the 9th ACM conference on computer and communications security. ACM Press, New York, pp 207–216

Girvan M, Newman MEJ (2002) Processing of the national academy of science of the United States of America. Community structure in social and biological networks 99:7821–7826

Golbeck J, Hendler J (2004) Inferring reputation on the semantic web. In: Proceedings of the 13th international world wide web conference

Golbeck J, Hendler J (2004) Accuracy of metrics for inferring trust and reputation. In: Proceedings of the 14th international conference on knowledge engineering and knowledge management

Goldberg D, Nichols D, Oki BM, Terry D (1992) Using collaborative filtering to weave an information tapestry. Commun ACM 35(12):61–70

Kamvar SD, Schlosser MT, Garcia-Molina H (2003) The EigenTrust algorithm for reputation management in P2P networks. In: Proceedings of the twelfth international world wide web conference

Langville AN, Meyer CD (2006) Google's page rank and beyond. Princeton University Press, Princeton

Passant A et al (2009) Enabling trust and privacy on the social web, W3C workshop on the future of social networking, 15–16 Jan 2009

Sakimura N et al (2011) OpenID connect basic client 1.0-draft 15. http://openid.net/specs/openid-connect-basic-10-15.html

Yamagishi T (1998) Shinrai no Kozo – Kokoro to Shakai no Shinka Gemu [Structure of reliability – evolutional games of mind and society]. University of Tokyo Press, Tokyo

Yamagishi T, Yoshikai N (2009) Net hyoban Syakai (Network Reputation Society). NTT press, ISBN978-4-7571-0266-8

Yamasaki S (2010) A dynamic trust estimation method for 'Persona' from the human relationship of social web. In: Proceedings of the 10th annual international symposium on applications and the internet (SAINT2010), pp 300–303

Chapter 6
Smartphones: The Next Generation Medication Administration Tool

Eizen Kimura

6.1 Introduction

A common aim of healthcare professionals is that patients regain a healthy state under their professional and dedicated treatment. However, the "To Err is Human" (Kohn et al. 2000) report released by IOM (Institute of Medicine) in 1999 emphasized the fact that many patients suffer from medical accidents every year. This report brought to the healthcare domain the concept of "Safety Management" as a medical accident measure, and attempted to ensure "Patient Safety" by the introduction of the ICT (information and communications technology) check system.

The Japanese Council for Quality Health Care reported in 2009 that nurses had the most frequent error incident reports, followed by physicians (Japan Council for Quality Health Care Division of Adverse Event Prevention 2012). Most incidents occurred in the patient's room (inpatients) and the operating room; the most frequent issues were "Bedside assistant jobs" and "Treatment care/procedure". The most frequent reported causes were "neglecting to check", "misjudgment", and "neglecting to observe." The report noted that the number of reported incidents had not changed since 2005. It was also noted that 41 % of medical errors ensued from medication administration (Lisby et al. 2005). Medical safety management in medical practice is still an ongoing issue.

The rising cost of medical technology development has encouraged national interest in "sustainable health services". For healthcare management reform, the focus on information gathering and analysis of EHRs (Electronic Health Records) is a pressing concern. We note the importance of medical safety practices, such as properly documenting medical records, tracking expenditures for hospital business analysis, and ensuring traceability and accountability.

E. Kimura (✉)

Department of Medical Informatics of Medical School of Ehime University, 791-0295 Situkawa Toon City, Ehime, Japan

e-mail: ekimura@m.ehime-u.ac.jp

S. Uesugi (ed.), *IT Enabled Services*,
DOI 10.1007/978-3-7091-1425-4_6, © Springer-Verlag Wien 2013

Medical practice entails forms of consumption and, in this context, the idea of building a mobile medication checking system has emerged. These mobile devices allow the user to take them away from the staff station, enabling the checking of medication order lists and consumed medications at the same time. However, there are many problems peculiar to the healthcare field; trials using mobile terminals with ICT support have a long way to go before being acceptable for medical use. We will give an outline of the background of BCMA (Bar-Corded Medication Administration) (Koppel et al. 2008), introduce some trials on it, and discuss the way forward.

6.2 Issues Relating to BCMA

6.2.1 Institutional Background

In Japan, "The Drugs, Cosmetics and Medical Instruments Act" and "The law to revise the drawing of blood and blood donation business control methods" came into effect in 2005 (Pharmaceutical and Food Safety Bureau 2004). Biogenous medical products, except plasma derivatives, require management of the production history. Not only is a product's history tracked during the initiation stage and during production but the selection of the donor (the source of raw materials) should also be able to be confirmed. In addition, it is necessary to maintain records of antipollution maintenance and manufacturing tracking.

MHLW mandates drug companies to keep detailed records of followup of the donor user, the infectious disease submission report, the mandatory display of primary material if a drug is biogenous product, and the monitoring of proper usage at the post market. Drug companies must make sure that all records are connected and verified to ensure accountability in case of future incident investigation.

Historical background of the constitution of the law was the experience of the lawsuit on adverse drug reactions and the investigation of the casual association between administered drugs and the adverse drug reactions had processed with difficulty because of lacking sufficient records. We recognize that extensive traceability is required in order to undertake root cause analysis into the harmful effects of drugs.

In September 2006, to prevent accidental misunderstandings on drug applications and to ensure traceability, guidelines for implementing medication barcodes were given in office notice No.0915001 "The barcode display on an ethical drug" by MHLW (Ministry of Health, Labour and Welfare) Pharmaceutical and Food Safety Bureau (2006). The ethical drugs were categorized into five kinds: biogenous products, specific biogenous products (among biogenous products, main raw material is from human blood or tissue), oral medicine, intravenous, and external medicines. Packaging schemes are categorized into three kinds: dosage packaging unit, sell packaging unit and bale packaging unit. The notice also requires manufacturers to provide serial number manufacturing codes, quantity, and expiration date on the barcode.

The JAN (Japanese Article Numbering) (The Distribution Systems Research Institute 2012) code is the product identification code of the Japanese Industrial Standards; it is designed to be compatible with EAN (European Article Number) of GS1. GS1 (Global Standard One) is a global organization dedicated to the design and implementation of global standards of supply and demand chain. The number of the packaging type and the JAN code together make up the product code of the ethical drug. The information level on a dosage packaging unit is necessary to ensure the proper medical treatment at the end point of accountability. "GS1 RSS (Reduced Space Symbology, recently renamed as "DataBar".) limited composition symbol CC-A" and "RSS-14 stack composition symbol CC-A" should be used.

As of 2010, the product code of the sale packaging unit is displayed on more than 99 % of medical supplies. However, as for the manufacturing number or the manufacturing code, only a specific biogenous product is 100 % (Ministry of Health Labour and Welfare Health Policy Bureau Economic Affairs Division 2012). The degree of labeling on other medical supplies is still low; the environment for traceability in the medical field is still developing. Neither barcodes nor RFID (Radio Frequency Identification) are completely adequate for labeling purposes at the moment; it is necessary to build a system that supports both barcodes and RFIDs.

6.2.2   Healthcare Management

The medical treatment system in Japan has traditionally used fee-for-service payments, but recently the payment method has been shifting to the prospective payment system. MHLW introduced the DPC (Diagnosis Procedure Combination) (Okamura et al. 2005) in 2003. While the DRG/PPS system is a "per-case payment" system, the DPC is "per-day payment"; it is generally believed that there is an incentive in shortening the average length of stay (LOS) to reduce medical expenses. When part of the medical treatment is fixed, controlling medical costs becomes the most important task for healthcare.

It is common that the business analysis in hospitals distributes and assesses the cost of ethical drugs based on the billing information of each patient and on a departmental basis. The practitioners enter the orders, and these orders are transferred to the medical business department to gather the claim information and generate the medical care claims bills.

The billing transaction process in Japan was computerized in 1999. However, the current ordering system merely records the history of ordering and dispensing medications, allows administrative checking, and converts the transaction information to billing claims. The billing rules are so complex that the medical treatments and the billings do not have a one-to-one relation. The billing does not reflect the actual consumption of ethical drugs for medical administration purposes. Thus, it is impossible to determine the accurate picture of healthcare expenditure strictly from the billings.

Adopting the unit control of medical supplies and recording the exact medical treatment and procedures will allow for real cost accounting. However, checking the unit controlled medical supplies by scanning barcodes may increase the burdens on healthcare practitioners. Personnel costs comprise nearly half of medical expenditures (Ministry of Health Labour and Welfare Health Policy Bureau Economic Affairs Division 2009); thus, increasing the burdens on practitioners will put pressure on management to refine business analyses and practices. A medication checking system must require the least amount of personnel labor possible.

6.2.3 Operating Issues with Regard to the Ordering System

Healthcare services around a Japanese medical ward are limited due to poor housing conditions (Ministry of Health Labour and Welfare 1995). The lack of resources makes it difficult to build a cost effective and timely delivery system of maintaining a stock of ethical drugs at the ward and distributing them efficiently. Common practice is to have the pharmaceutical department arrange the delivery date for the drug order and deliver the drugs at the appointed time. The ordering system is recognized as a tool for reserving or backordering drugs over a long period (Tomohiro 2005). The real computer-based injection designation system has recently emerged.

Tanaka et al. reported that among all intravenous infusions at the hospital in 2007, less than 1 % of treatments were not authenticated, due to an emergency order by a physician (Kaihara et al. 2009). Among authenticated orders, 55 % were changed or canceled. This study shows that even when put on alert, nurses have tendency to repeatedly enforced some orders to be authenticated. The fact that 4 % of the override actions were the cause of malpractice incidents indicates that the healthcare workers are too close to medical accidents. Medical treatment cannot always be scheduled. A failsafe design in the checking system at a ward is very important. In addition, the long time span between dispensing drugs and administering treatment can also cause frequent order changes and cancellations.

Lacking a supply system that supports sustainable and prompt delivery for urgently required ethical drugs, practitioners have a tendency to order extra drugs so as to have them on hand. To change the focus from ordering resources to indications for medical treatment, the system should shorten the time between dispensing a drug and administering it. Collecting the log of authentication, the list of medicated ethical drugs, and the log of treatment time will enable realtime stock forecasting and better stock control.

6.2.4   Securing Medical Safety

To ensure medical safety, BCMA helps practitioners to ensure the five "rights" of medication administration (Koppel et al. 2008): "right patient," "right drug," "right dose," "right route," and "right time." On the other hand, there is a report pointing out that the effect of BCMA is limited and may even be one of the causes of medical accidents (Koppel et al. 2008). The controversy over the usefulness of BCMA stems from a lack of standardized definitions relating to the information and the workflow that BCMA should handle (Akiyama and Atsushi 2009; Henneman et al. 2007). In that sense, BCMA is not a mature system. There have also been few discussions about the absolute merit of the five "rights" (Shane 2009); the information for analysis and authentication is still not fully elucidated.

Akiyama et al. suggested that authentication of the "right drug" not only implies authentication of the right kind of drug, which is exactly what the physician has prescribed, but also implies checking whether a drug to be co-infused is co-infused correctly, as well as accessing a drug adverse reaction database to ensure that these drugs have no side effects related the medical administration (Akiyama et al. 2008a). After drugs are co-infused, there is a decrease in drug efficacy and bacterial growth as time goes on; mixing drugs more than 1 h from the prescribed time is inappropriate (Schneider et al. 1998). "Right route" is commonly focused on confirming the intravenous injection route. Additionally, improper treatments that deviate from the scheduled time should be suppressed. Checking whether additional treatments are medically appropriate is also required; this is conducted by calculating the daily dose (Akiyama and Atsushi 2009). In other words, securing medical safety does not involve merely authenticating items and times; it also requires that treatments are performed appropriately, using the right medications, at the prescribed time and place, and properly conducted using the correct workflow. To that end, we have to develop a system that accesses an adverse event database and queries the clinical decision support system instantly, via an authentication process, to judge whether the request is correct according to the "five rights".

6.2.5   Information Security

In healthcare, the sharing of patient information between practitioners, as well as immediate access to the information, should be guaranteed. Without this guarantee, patients' lives may be endangered. Efficient medical treatment may be hampered if access control with regard to medical records is inappropriate. The healthcare domain should establish an original security policy that balances the protection of information, considering fundamental patients' rights, and the disclosure of patient information to practitioners to ensure medical safety (Tsukuma et al. 2001).

Some paper suggested that instead of managing access control strictly, we should add audit functions to the medical information system. We may enhance accountability by tracking personnel access and preventing fraudulent activity. In other words, guaranteeing immediate access to patient information, determining who accesses it, and recording these logs in a tamper-proof format are more important than preventing illegal access to patient information in healthcare.

6.2.6 Ethical Drugs and Traceability

Traceability in the healthcare domain is the ability to track ethical drugs from production to patient use. To that end, product information should accompany medical supplies from production to distribution. However, barcode re-labeling may be performed during physical distribution, hence it may not be possible to locate critical information, such as the lot number, and a medical center may also get the wrong ethical drugs if the re-labeling process itself is done in error.

One problem in this regard is that the information management units are different between the production, physical distribution and dosage stages. To bridge the gap between information management units, barcode re-labeling is undertaken. The production phase and the dosage phases require "Unit Dose", but in the physical distribution phase, labeling is done according to various packaging forms, such as palette, lot, the cardboard box, etc. Every time drugs are packaged and unpackaged, re-labeling may be required.

GS1 (Association of the old international EAN) has standardized the physical distribution process to resolve these issues. In the GS1 standard, GS1-128 (UCC/ EAN-128), RSS (Reduced Space Composite Symbology) and RFID are used as Data Carriers. GTIN (Global Trade Item Number) and SSCC (Serial Shipping Container Code) are approved as the standard data format. GTIN focuses on consumption while SSCC focuses on physical distribution; GTIN and SSCC are mutually interchangeable. In the healthcare domain, most pharmaceutical companies use the EPC Global code, and it has become the de facto standard. The EPC uses SGTIN (Serialized GTIN) that has interchangeability with GTIN. In Japan, JAN (Japanese Article Number) is specified based on EAN and also has interchangeability with GTIN.

Japan has mandated that all ethical drugs are to be labeled with barcodes after 2008. Unit item management can identify the kind (name) of ethical drugs, but cannot identify individual ampoule bottles. Single item management can use the serial number to distinguish individual products, and it is indispensable for ensuring medical safety. Accidents in which practitioners mistake the mixed injection drug for another because the bottles appear the same will be avoided by distinguishing the individual bottles (Kondoh et al. 2008).

Therefore, SGTIN enables single item management and is regarded as an important component in the medical field. However, as the barcode is not currently displayed on all medical supplies, complete source marking is not now available. Every medical institution has to apply private "in-house" markings in each medical institution, which creates additional administrative burdens.

6.3   Issues Relating to BCMA

Nurses in a ward carry out most of the orders for patients during a hospitalization. Blood transfusions are primarily reported by the nurses at a ward, implying that many invasive and high-risk medical treatments are conducted at the ward; dependable authentication should be performed beforehand. Below is a typical workflow process from ordering the blood transfusion to the actual transfusion to give a general idea of the workflow at a healthcare setting:

A physician inputs the blood transfusion order via CPOE (Computer-based Provider Order Entry). The blood type of the patient is registered on an electronic medical record, previously inspected by a blood transfusion department system. The blood transfusion department accepts the order from the physician and performs the radiation on the blood bag and dispenses it. At the staff station, nurses verify the dispensed blood with the form issued by CPOE; then, they carry the blood to the patient's bedside. The nurse scans the barcode on the staff identification plate and inputs it as the performer's identity. Then she scans the barcode on the hospital card of the patient. Finally, she scans the barcode on the blood bag to perform three points authentication (Dzik 2007).

Because the authentication process requires the authenticating person and materials at the ward, mobile terminals such as a PDA (Personal Digital Assistant) and PHS (Personal Handy-phone System), attached to a barcode reader, are widely used. In Japan, BCMA has been implemented on mobile terminals since the 2000s and has spread rapidly (Akiyama 2007a). The number of negative incidents has been remarkably reduced in hospitals using BCMA (Watanabe et al. 2006; Matsuda et al. 2006; Makoto 2004); however, such incidents have not been completely eliminated.

The patients wear a wristband with a barcode in many hospitals, but the responses from patients in this regard have not been positive (Yano et al. 2008). Therefore, some hospitals use a hospital card instead of the wristband, using the wristband only in a limited number of situations, such as during operations.

However, these sorts of measures may be the cause of certain problems. Too many barcodes required for authentication (Tomohiro 2005), and the complicated operation of the mobile terminal, can lead to nurses bypassing authentication (Watanabe et al. 2006) or refraining from using BCMA; there are many reports that nurses have been hesitant in using BCMA (Yano et al. 2008; Tateishi et al. 2008). For example, the light of a scanner is very bright, which can disturb sleeping patients. Another example sets out how a nurse gave up using the terminals because the terminals were not charged and spare terminals could not be found immediately.

A system that reduces the burden on practitioners is, obviously, desirable (Lisby et al. 2005; Kaplan 2005). Koppel analyzed the 'workarounds' of BCMA and their causes in detail (Koppel et al. 2008): (1) The user administers medication without scanning the medication barcode to confirm whether it is the correct medication, leading to wrong medication administered, (2) the user administers medication without scanning patient ID barcode to confirm it is the correct patient, leading to the wrong patient receiving medication, and (3) the user does not check or verify the patient's new medication orders before administering medication, leading to wrong medication dosage or administration route. The most prominent probable causes are: (1) The scanning procedure is slower or more difficult than other methods because it may conflict with workflow efficiency (Watanabe et al. 2006; Tateishi et al. 2008), (2) Users are not well trained in using the adequate procedures, and (3) Users are not aware of hospital medication use policies, e.g., double checking for high-risk medications. In other words, from the viewpoint of BCMA, practitioners act inappropriately; from the standpoint of human design, current BCMA may be one of the worst products in the medical field, due to a lack of sensitivity about human nature built into its design.

The PDA introduced from 2000 to 2007 cannot be said to be comfortable for the nurse because its performance (Watanabe et al. 2006), size and weight are far behind the ideal form (Yano et al. 2008). Windows CE-based terminals were mainstream in those days, and these terminals had to expand to contain the high-capacity battery and scanner modules needed to support longtime operation in the ward. As a result, the size and weight of the terminal were increased to the point where it was difficult to call it "mobile". This became a major reason why nurses did not want to effectively use BCMA. Other reasons to avoid BCMA include the fact that the barcode is sometimes hard to scan, depending on the surface shape (Yano et al. 2008; Tateishi et al. 2008); additionally, the pattern of the barcode may be distorted, broken, or stretched.

Currently, most healthcare settings do not use the barcode labeled at the time the product is shipped from the manufacturer, but use a barcode that the SPD (Supply Processing Distribution) center or pharmaceutical department provides when it is delivered to the healthcare setting. When drugs require refrigeration, the labels may freeze and come off or condensation may blur the information (Yano et al. 2008). When the labels have to be re-labeled, there may be a chance that the wrong label is applied. Some healthcare facilities re-label drugs after mixing an injection at a ward. There may be also a possibility of mislabeling when the practitioner mixes drugs for multiple patients at the same time.

6.3.1 RFID Overcomes the Problems of Barcode

It has been repeatedly pointed out that barcode authentication, relying on optical technology, has usability problems: (1) A scanner's orientation and focus need to be precisely oriented to the barcode, which requires great care, (2) Scratches, dirt, or leaking fluids, such as blood, can make the barcode unreadable, (3) The barcodes on a soft and deformable surface such as a wristband or intravenous feeding bags are sometimes unreadable, (4) Attaching barcodes to a patient's body may disturb the patient's sleep and rest. It may also upset patients because it detracts from their individuality, (5) It requires scanning multiple barcodes to mix multiple injections, (6) With traditional distribution processes, multiple barcode standards coexist and sometimes re-labeling is required (Kondoh 2007). These defects place a burden on practitioners.

In comparison to optical barcode technology, the RFID (Radio Frequency Identification) utilizes the characteristics of radio, which has the following advantages: (1) The direction of an RFID reader does not have to exactly match the IC tag, and it is readable even if there is shielding (noncontact authentication), (2) RFIDs can read more information and more precisely than a barcode. Some IC tags can also be written to or have write-once capability, (3) Multiple IC tags can be read at the same time, (4) There is enough room in RFID for reading information so automation of scanning materials is possible. In other words, the problems of barcodes will be resolved by utilizing RFID's characteristics.

6.3.2 RFID Trials in Healthcare

Kondo et al. developed a medical PDA with RFID reader and a wristband that has an IC tag built-in (Kondoh 2007). They conducted a feasibility study at a medical ward in 2004. Before medical treatment, the practitioners perform user authentication by scanning the IC tag in a staff identification card. Then, they scan the tag on the medication and the patient's wristband to confirm that the drug matches the order of the physician. As a result, injection related incidents decreased, and they can expand the coverage to both blood transfusions and outpatient chemotherapy. There is the case replacing the conventional barcode with IC tags. This study showed that an IC tag has a higher input efficiency and more efficient treatment than the cases with barcodes (Kondoh 2007; Ota et al. 2008).

One in 10,000 people are reported to have surgical instruments or gauze left inside the body after an operation (Gawande et al. 2003). The number of surgical instruments used in one operation can be several hundred, depending on the operation. The nurse in an operating room has to count all instruments and to confirm that no items are missing before, during, and after the surgery. The surgical instruments are stored in a container and sterilized after irrigation. The distribution of instruments to each container is also performed by manual labor, and approximately 2 % of distribution work may include the wrong items (Yamashita et al. 2009).

The MHLW (Ministry of Health, Labour and Welfare) mandates, via the Amendment of Pharmaceutical Affairs Law of 2007, the safe management of surgical instruments and setting the expiration date of instrument based on usage count (Ministry of Health Labour and Welfare Pharmaceutical and Food Safety Bureau Safety Division 2003). This mandate aims to establish more effective medical safety practices by changing maintenance management from a non-binding guideline to a requirement (Akiyama 2007b). However, a medical institution with many surgeries often has 4,000–7,000 operations every year and manages more than 100,000 instruments (Yamashita et al. 2009). At present, no healthcare facility has managed surgical instruments on a unit management basis. There is no evidence available for surgery frequency and the failure rate of instrument counting.

MHLW's demands for safety management will result in a huge burden on healthcare unless countermeasures are taken.

The use of IC tags with surgical instruments is expected because of the IC tag's capability of providing multiple item authentications at the same time. Marcario had tried to find the gauze left in a human body by scanning a small IC tag attached into the gauze (Macario et al. 2006). However, we have to overcome a weak point of the IC tag in order to account for surgical instrument materials. Surgical instruments are mainly made of metal, and sterilization is necessary. Ethylene oxide gas sterilization is used for precision instruments, such as an endoscope, that should not be heated at high temperatures. Steel devices, such as scalpels or forceps, are sterilized by high-pressure steam (Wolfgang et al. 2009). IC tags need to be resistant to high temperatures and chemical agents. IC tags should also have impact resistance because they can be subjected to the impact of automated cleaning and transportation. IC tags are required to maintain readability under these severe environments.

To address these issues, Yamashita et al. developed the ceramic IC tag, which buried a small IC tag in a ceramic body; this allows the tag to perform durability management by logging usage frequency and the history of surgical instruments through the cycle of irrigation, sterilization, and keeping (Yamashita et al. 2007). The ceramic IC tag is resistant to high temperature, pressures, and chemical agents. The advent of the ceramic IC tag eliminates the areas where IC tags cannot be used in healthcare, and may lead to ubiquitous use. The issues that remain to be solved are reducing the costs, standardization of the information on IC tags, and development of a tag management system.

6.4 Challenges for the Future

6.4.1 Workarounds for Barcode

There are some problems with regard to BCMA that need to be resolved. One is that the mobile terminals of BCMA have a tendency to be heavy, because a barcode scanner module has to be attached to the terminal. The labels on a deformable package or on a curved surface are difficult to read. Machine error, unreadable barcodes, and users forgetting to read the barcodes are major risk factors that threaten medical safety (Koppel et al. 2008). However, barcode systems cost less than RFIDs, hence it is expected that barcodes will coexist with RFID in the future.

One future challenge is that of developing a mobile terminal that can process barcodes more easily. One possible solution is a smartphone equipped with an AF (Auto Focus) camera. Smartphones have high-speed processors so they can handle more advanced image processing than conventional PDAs, in a shorter amount of time. Reading barcodes with AF cameras and performing advanced image processing may increase the recognition rate of barcodes. As robust barcode recognition algorithms are researched (Gallo and Manduchi 2009; Wachenfeld et al. 2008), it is expected that more efficient BCMA may be developed.

6.4.2   Expand the Use of RFID in Healthcare

Investigation of the effects of electromagnetic waves from RFID devices in healthcare settings has not been adequately performed (Akiyama et al. 2006). There have been studies of the effect of radio waves produced from RFID devices on medical equipment in laboratory settings (Ministry of Internal Affairs and Communications 2006), but little research has been conducted in real medical facilities.

There are huge amounts of transaction data that contain patient status, medication history, and the list of administered drugs with a lot-number. A distributed processing technology to handle such data efficiently is required to realize realtime CDSS (Clinical Decision Support System). To effectively work with the life cycle management of ethical drugs' supply, safekeeping, mixture, administration and disposal, the standardization of a drug master code and a medical information model should be developed. Solving these issues will accelerate the expansion of RFID in healthcare.

6.4.3   Data Mining for Medical Safety

Akiyama et al. developed POAS (Point of Act System), a system that collects the transaction logs of medication administered to patients. By conducting data mining on tens of millions of transaction data, they found that warnings about an injection occurred most frequently when the nurses' shift change occurred (Akiyama et al. 2008b). Data mining based on the real dynamics of medical treatment is impossible with conventional ordering systems and medical electronic records. By using mobile terminals, the practitioner, patient, medications, injections, place, time, and routes are all recorded, allowing investigation into which part of the workflow medical accidents most frequently occur. Analyzing the time lag and the route between ordering and administering medication may optimize stock and material distribution.

This kind of data mining has already been performed in other fields, but data mining in healthcare is still in the early phases of development. As I have described previously, the standardization of the medical information model is still developing, and we have to combine complicated, structured data to build data mining targets because the treatment process includes many kinds of administering medication, and these administration transactions are stored separately.

6.4.4 Introduction of Smartphone to Healthcare

The introduction of information technology to the healthcare field has demonstrated that such technology can reduce medication errors; however, there is no evidential support for increased adherence to protocol-based care (Wu et al. 2006), and many studies lack a financial analysis of the introduction of mobile terminals in preventing medical accident; the real cost effectiveness is still unclear. However, making the shift to smartphones from PDAs or special-purpose barcode readers will improve the cost-effectiveness aspect.

PHS (Personal Handy-phone System) is very common in Japanese healthcare settings; the GSM (Global System for Mobile) and CDMA (Code Division Multiple Access) based cellular phones are not used. The evidence of the influence on medical equipment of cellular phone's radio waves has been unclear until recently, so hospital managers have hesitated to introduce GSM or CDMA cellular phones. The smartphone's capability to connect to multiple channels such as Wi-Fi, 3G and 4G wireless technology is an advantage, but its advantage also may be a weakness because more channels means it is more open to security risks.

Through the smartphone, enormous amounts of personal information will be exchanged. We have to develop distributed processing systems that can service the realtime clinical decision support while providing adequate security (Ministry of Health Labour and Welfare 2010). It will be necessary to develop the authentication system closely with cloud computing technology. There are many issues that hospital managers should take into consideration before introducing smartphones into medical institutions.

6.5 Conclusion

Research into ensuring medical safety and analyzing the medical business process with mobile terminals has been conducted for more than 10 years. The course of this research has not been smooth, and there are many remaining problems. However, I expect that the advent of smartphone implementation will provide a break-through for several reasons: The major problems of BCMA are the low-quality design of user interface and the performance issue on barcode reading. Smartphone processor performance is now surpassing former high-end workstations while the more sophisticated interface and more robust barcode recognition by high-performance image processing will make BCMA more useful.

Furthermore, a smartphone equipped with NFC (Near Field Communication) will be expected to integrate barcode and RFID into one item. This capability will be available in many commercial products so that all healthcare members may change from PHS to smartphones, and they will be ready to conduct medical administration authentication no matter what the time or location. In addition, smartphones support many channels such as Wi-Fi, 3G and 4G wireless technologies at the same time. The practitioners can conduct medical administration anywhere.

The smartphone seems promising, but there is little empirical evidence of its effectiveness in the healthcare domain. We have to verify the smartphone's usability in the future.

References

Akiyama M (2007) Risk management and measuring productivity with POAS – point of act system – a medical information system as ERP (enterprise resource planning) for hospital management. Methods Inf Med 46(6):686–693

Dzik WH (2007) New technology for transfusion safety. Br J Haematol 136(2):181–190

Pharmaceutical and Food Safety Bureau (2004) Enforcement of Law Partially Revising the Pharmaceutical Affairs Law and the Blood Collection and Donation Service Control Law

Fuchs W et al (2009) Proper maintenance of instruments. Instrument Preparation Working Group

Gallo O, Manduchi R (2009) Reading challenging barcodes with cameras. Proc IEEE Workshop Appl Comput Vision 2009(7–8):1–6. doi:10.1109/WACV.2009.5403090

Gawande AA et al (2003) Risk factors for retained instruments and sponges after surgery. N Engl J Med 348(3):229–235

Henneman EA et al (2007) Increasing patient safety and efficiency in transfusion therapy using formal process definitions. Transfus Med Rev 21(1):49–57

Japan Council for Quality Health Care Division of Adverse Event Prevention (2012) Project to collect medical near-miss adverse event information 2009. Annual report 2009, Feb 2012. http://jcqhc.or.jp/

Kaplan H (2005) Getting the right blood to the right patient: the contribution of near-miss event reporting and barrier analysis. Transfus Clin Biol 12(5):380–384

Katsuyuki K (2007) IC-tags with hospital information system. J Infrom Process Soc Jpn 48 (4):338–343

Katsuyuki K et al (2008) Ubiquitous technology as medical and social infrastructure. In: The 28th joint conference on medical informatics, vol 28(Suppl.), pp 60–66

Kazuhiko Y et al (2007) IC tag for information management of surgical instruments on operating room. J Inform Process Soc Jpn 48(4):349–353

Kazuhiko Y et al (2009) Development of management system of surgical instruments by RFID for improvement in the quality of operation. The Jpn J Med Instrum 29(suppl):369–374

Kazutika W et al (2006) Developing nurse care aid terminal with web service technology. In: The 26th joint conference on medical informatics, vol 26, pp 1283–1284

Kohn LT, Corrigan J, Donaldson MS (2000) To err is human: building a safer health system. The National Academies Press, Washington, DC. ISBN 0309068371

Koppel R et al (2008) Workarounds to barcode medication administration systems: their occurrences, causes, and threats to patient safety. J Am Med Inform Assoc 15(4):408–423

Kuroda T (2005) Introducing ubiquitous nursing support system for Kyoto University Hospital. New Med Jpn 368:176–180

Lisby M, Nielsen LP, Mainz J (2005) Errors in the medication process: frequency, type, and potential clinical consequences. Int J Qual Health Care 17(1):15

Macario A, Morris D, Morris S (2006) Initial clinical evaluation of a handheld device for detecting retained surgical gauze sponges using radiofrequency identification technology. Arch Surg 141 (7):659

Masanori A (2007) Traceability and management in medical field: according to an international trend. Jpn Soc Med Instrum 77(6):372–380

Masanori A, Atsushi K (2009) The 5 rights on IT contributes to patient safety. J Jpn Assoc Med Inform 29(Suppl):784–787

Masanori A et al (2006) Feasibility study for RFID tag effectiveness in medical field-special coordination funds for promoting science and technology of the ministry of education, culture, sports, science and technology, the Japanese Government. In: The 26th joint conference on medical informatics, vol 26(Suppl.), pp 146–149

Masanori A, Hajime N, Akihiko S (2008a) Feasibility study of traceability in medical field with international standard RFID tag. In: The 28th joint conference on medical informatics, vol 28 (Suppl.), pp 496–499

Masanori A et al (2008b) Data mining analysis with electronic medical record for risk management. In: The 26th joint conference on medical informatics, vol 25(Suppl.), pp 370–373

Ministry of Health Labour and Welfare (1995) Annual report of ministry of welfare 1995 (1995 Feb 2012). http://wwwhakusyo.mhlw.go.jp/wpdocs/hpaz199501/b0028.html

Ministry of Health Labour and Welfare (2010) Safty management guideline for medical information system Ver. 4.1. (2010 Feb 2012). http://www.mhlw.go.jp/shingi/2010/02/dl/s0202-4a.pdf

Ministry of Health Labour and Welfare Health Policy Bureau Economic Affairs Division (2009) Healtcare management control indicator (2009 Feb 2012). http://www.mhlw.go.jp/topics/bukyoku/isei/igyou/igyoukeiei/keieisihyou/21kannri.html

Ministry of Health Labour and Welfare Health Policy Bureau Economic Affairs Division (2012) The report on progress of the informalization for ethical drugs (2012 Feb 2012). http://www.mhlw.go.jp/stf/houdou/2r985200000157qp.html

Ministry of Health Labour and Welfare Pharmaceutical and Food Safety Bureau Safety Division (2003) Voluntary inspection of orthopedic surgical instruments. Notification No 0311001. 3 Mar 2003

Ministry of Health Labour and Welfare Pharmaceutical and Food Safety Bureau Safety Division (2006) Notification No.0709004 of the PFSB Enforcement of barcode display on the ethical drugs

Ministry of Internal Affairs and Communications (2006) Investigation into about the influence of electronic wave on implantable medical equipment: findings on cell-phone (1.7 GHz zone) and electronic wave of electronic tag system (950 MHz zone) (cited Feb 2012). http://www.tele.soumu.go.jp/j/sys/ele/medical/cyousa/index.htm

Norihiko T et al (2008) Improvement of the nursing support system by using personal digital assistant. In: The 26th joint conference on medical informatics, vol 26, pp 961–962

Ohara M (2004) Safety management system in national center for child health and development. In: The 24th joint conference on medical informatics, vol 24, pp 137–139

Okamura S, Kobayashi R, Sakamaki T (2005) Case-mix payment in Japanese medical care. Health Policy 74(3):282–286

Rie M et al (2006) The effect of wrist-band risk management system operating for 2 years. In: The 25th joint conference on medical informatics, vol 25, pp 366–367

Sakiko O et al (2008) Comparison of RFID and bar code system for support of medical practice in the ward. In: The 28th joint conference on medical informatics, vol 28(Suppl.), pp 679–681

Schneider MP, Cotting J, Pannatier A (1998) Evaluation of nurses' errors associated in the preparation and administration of medication in a pediatric intensive care unit. Pharm World Sci 20(4):178–182

Shane R (2009) Current status of administration of medicines. Am J Health-Syst Pharm 66(5 suppl 3):42–48

Shigekoto K et al (2009) Information technology applied for the supply chain of drugs and equipments in hospitals, especially aiming at patients' safety. In: The 29th joint conference on medical informatics, vol 29(Suppl.), pp 300–302

The Distribution Systems Research Institute (2012) Japanese (Article number code. Feb 2012). http://www.dsri.jp/jan/

Tsukuma H et al (2001) Security policy for management of electronic medical record system to guarantee the patient's rights. In: The 21st joint conference on medical informatics, vol 21, pp 717–718

Wachenfeld S, Terlunen S, Jiang X (2008) Robust recognition of 1-d barcodes using camera phones. In: 19th international conference on pattern recognition (ICPR 2008), Tampa, pp 1–4

Wu S et al (2006) Systematic review: impact of health information technology on quality, efficiency, and costs of medical care. Ann Intern Med 144(10):742–752

Youko Y et al (2008) Effects and problems of PDA with tag leader on the inpatient management. In: The 28th joint conference on medical informatics, vol 28, pp 673–678

Chapter 7
Weaknesses of the E-Government Development Index

Eltahir Kabbar and Peter Dell

7.1 Introduction

Although initially conceived in the relatively narrow context of Business Process Outsourcing (BPO), particularly to offshore locations, IT-enabled Services (ITeS) has matured and increasingly emphasizes innovation of conventional services (Uesugi 2008). An example of this in practice is the growth of e-government, in which conventional government services are provided in online environments.

However, as Weerakkody et al. (2009: iv) note, efforts to transform the public sector have largely resulted in "reinforcing old practices", rather than the innovation ITeS increasingly supports. For IT to fulfill its potential to enable true transformation of government, attention must be paid to the new and innovative services it enables. Thus, ITeS lie at the core of transformational government by allowing new practices, rather than simply automating inefficient and undesirable old ones.

E-government can promote economic development and encourage participation in service delivery processes (United Nations 2012). Consequently, there have been a number of efforts to monitor the state of e-government development in countries worldwide, the most widely reported of which is the United Nations E-Government Survey. This survey produces a regular series of rankings of countries' e-government development, based on a metric known as the UN E-Government Development Index (EGDI).

The importance of these rankings cannot be underestimated. They are widely cited and discussed, perhaps not least because as a product of the UN they are highly regarded by default. Yet there has been little critical examination of the index itself. One notable exception is the identification of statistical flaws in the manner in which the index is calculated (Whitmore 2012). However, even were this

E. Kabbar (✉) • P. Dell
School of Information Systems, Curtin University, Perth, Australia
e-mail: ekabbar@hct.ac.ae; P.T.Dell@curtin.edu.au

S. Uesugi (ed.), *IT Enabled Services*,
DOI 10.1007/978-3-7091-1425-4_7, © Springer-Verlag Wien 2013

to be rectified in future versions of the index there still remain questions as to how the index should be interpreted. It is not necessarily the case that an 'absolute value' of e-government development, which the UN metric provides, is particularly helpful if one's objective is to understand how well a country performs relative to its ability to do so; in such cases a metric that takes into account economic and sociocultural factors would be more appropriate.

With these thoughts in mind, this chapter examines the EGDI, and in particular its correlation with economic and sociocultural factors. We note that past literature in e-government has focused primarily on the supply side of e-government, including models of e-government evaluation and practices (Reddick 2004; West 2004); effectiveness of implementation and challenges of e-government services (Jaeger and Thompson 2003); success factors and implementation of E-government initiatives (Jaeger 2003; Traunmüller and Wimmer 2003). Very little attention has been paid to demand aspects of e-government. This chapter avoids this weakness by looking at both supply and demand sides. First, the following section examines macroeconomic issues that affect a country's ability to effectively deploy e-government services in the first place. Second, the chapter will examine some of the sociocultural factors that influence the population's willingness to use those services, and thus examines the demand side. Finally, the chapter concludes with some recommendations for practice and research.

7.1.1 The United Nations E-Government Development Index

The EGDI is calculated from data collected in the UN E-Government Survey. The most recent version of this EGDI was published in 2012 and is described in the UN documentation as "measuring the willingness and capacity of national administrations to use information and communication technology to deliver public services". The overall metric is based on three individual components. First, the Online Service Index measures the maturity of a country's e-government websites, such as their national website and related portals, and related websites from ministries such as education, labour, social services, health, finance and environment. Second, the Telecommunication Infrastructure Index derives a score for a country's telecommunications infrastructure based on five indicators: the proportion of Internet users, fixed telephone lines, mobile subscribers, fixed Internet subscriptions, and fixed broadband facilities. Finally, the Human Capital Index is calculated based on measures of a country's adult literacy and education enrolments.

Thus, the EGDI purports to take into account not only the maturity of e-government services in individual countries, but the capacity of those countries to deliver e-government services both in terms of the telecommunications infrastructure and the skills of the population to provide and exploit them.

Although the EGDI is statistically flawed, a more statistically valid ranking technique has not been widely adopted. Although other ranking methods have been proposed, the EGDI is the current standard ranking technique and – for better or for worse – public policy debate around the world is influenced by the EGDI more than any other measure.

It is prudent, therefore, to examine the EGDI in more detail to reveal possible consequences of its use. Given that it is likely to continue to remain the dominant measure of e-government success for some time, this will allow for public policy debate to ensure a wider view of the state of affairs of e-government development. This chapter examines economic and sociocultural factors related to the EGDI, details of which are explored in the following sections.

7.2 Economic Factors

Perhaps not surprisingly, the EGDI is strongly correlated to per capita GDP. Indeed, the UN E-Government Survey 2012 notes numerous associations between per capita GDP and the state of e-government in various countries. Nevertheless, the rankings presented in the UN report do not take income into account and the results are similar to a ranking of per capita GDP.

Therefore, in this chapter we present an alternative ranking based on the UN data and weighted to take into account per capita GDP, revealing somewhat unexpected results. We conclude with discussion of some of the possible consequences of an index which does not take these into account.

7.2.1 Incorporating GDP into EGDI

Using 2011 GDP data from the International Monetary Fund, a strong correlation was observed between each of the indices and per capita GDP. In the case of the Online Service Index (OSI):

$$OSI = 0.1352 \ \ln(GDP) - 0.7557; \ R^2 = 0.5219 \tag{7.1}$$

In the case of the Telecommunications Index (TI):

$$TI = 0.1698 \ \ln(GDP) - 1.1917; \ R^2 = 0.7424 \tag{7.2}$$

And in the case of the Human Capital Index (HCI):

$$HCI = 0.1257 \ \ln(GDP) - 0.4042; \ R^2 = 0.6023 \tag{7.3}$$

Finally, the overall EGDI is inevitably correlated with GDP, derived as it is from the three components above:

$$EGDI = 0.1436 \ \ln(GDP) - 0.7839; \ R^2 = 0.7712 \tag{7.4}$$

These correlations make intuitive sense: of course wealthy countries are more likely to have the ability to invest in online services, telecommunications infrastructure and education. However, the EGDI does not provide an easy basis for comparisons between countries that are not roughly equivalent in terms of GDP. For example, Qatar, the world's richest country in terms of per capita GDP, has an EGDI of 0.6405. In contrast, Zimbabwe, one of the world's poorest countries, has an EGDI of only 0.3583. Is one therefore to conclude that Qatar is "better" at delivering e-government than Zimbabwe? Or more efficient? Or something else entirely?

To resolve this difficulty we calculated the scores one would expect of each country based on per capita GDP, termed here Expected EGDI (EEGDI). These scores were calculated using the regression equation for overall EGDI above.

In order to determine if a country's actual performance as assessed by the UN (EGDI) was above or below their expected performance based on GDP (EEGDI) we then calculated a calculated the difference between these two scores. As this score is a reflection of a country's e-government development performance relative to what would be expected by GDP, we term this E-Government Development by GDP (EGDGDP):

$$EGDGDP = EGDI - EEGDI \qquad (7.5)$$

EGDGDP has a theoretical range of -1 to $+1$. In the best possible case a country has an EGDI of 1 but an EEGDI of 0, resulting in an EGDGDP of $+1$. On the other hand, in the worst possible case they have an EGDI of 0 and an EEGDI of $+1$, resulting in an EGDGDP of -1. A summary of the original EGDI scores and these differences is presented in Table 7.1.

Analysis of these EGDI differences reveals some very interesting – and unexpected – results. When per capita GDP is taken into account Zimbabwe would only be expected to have an EGDI score of 0.0981, yet that country's actual score of 0.3583 is much higher. On the other hand, Qatar is expected to score 0.8791, but has an actual score of only 0.6405.

By providing an indication of whether a country is performing above or below expectations, these differences are arguably more useful than the "raw" scores that are published by the UN; we therefore consider these to be a better measure of E-Government Development Performance than raw EGDI on its own.

7.2.2 Implications of Considering Economic Factors

Analysis that only considers the raw EGDI values and does not take into account the level of wealth in a country can be misleading. One such example can be drawn from the UN report itself, which describes the United Arab Emirates (UAE) as a "best practice case" (p. 23). *Prima facie* the UAE does look like a good example; the EGDI having increased from 0.5728 in 2008 to 0.7344 in 2012.

Table 7.1 E-government development index (EGDI), expected EGDI and e-government development weighted by GDP (EGDGDP)

Country	EGDI	EEGDI	EGD-GDP	Country	EGDI	EEGDI	EGD-GDP
Afghanistan	0.1701	0.2373	−0.0672	Burkina Faso	0.1578	0.2556	−0.0978
Albania	0.5161	0.5210	−0.0049	Burundi	0.2288	0.0845	0.1443
Algeria	0.3608	0.4901	−0.1293	Cambodia	0.2902	0.3375	−0.0473
Angola	0.3203	0.4562	−0.1359	Cameroon	0.3070	0.3358	−0.0288
Antigua and Barbuda	0.6345	0.6381	−0.0036	Canada	0.8430	0.7380	0.1050
Argentina	0.6228	0.6208	0.0020	Cape Verde	0.4297	0.4142	0.0155
Armenia	0.4997	0.4529	0.0468	Chad	0.1092	0.2882	−0.1790
Australia	0.8390	0.7385	0.1005	Chile	0.6769	0.6090	0.0679
Austria	0.7840	0.7441	0.0399	China	0.5359	0.5136	0.0223
Azerbaijan	0.4984	0.5419	−0.0435	Colombia	0.6572	0.5398	0.1174
Bahamas, The	0.5793	0.6988	−0.1195	Comoros	0.2358	0.2236	0.0122
Bahrain	0.6946	0.6704	0.0242	Congo, Democratic Republic of	0.2280	0.0662	0.1618
Bangladesh	0.2991	0.3067	−0.0076	Congo, Republic of the	0.2809	0.4218	−0.1409
Barbados	0.6566	0.6627	−0.0061	Costa Rica	0.5397	0.5727	−0.0330
Belarus	0.6090	0.5959	0.0131	Côte d'Ivoire	0.2580	0.2814	−0.0234
Belgium	0.7718	0.7312	0.0406	Croatia	0.7328	0.6299	0.1029
Belize	0.3923	0.5230	−0.1307	Cyprus	0.6508	0.6881	−0.0373
Benin	0.2064	0.2778	−0.0714	Czech Republic	0.6491	0.6753	−0.0262
Bhutan	0.2942	0.4665	−0.1723	Denmark	0.8889	0.7284	0.1605
Bolivia	0.4658	0.4374	0.0284	Djibouti	0.2228	0.3379	−0.1151
Bosnia and Herzegovina	0.5328	0.5116	0.0212	Dominica	0.5561	0.5838	−0.0277
Botswana	0.4186	0.5955	−0.1769	Dominican Republic	0.5130	0.5381	−0.0251
Brazil	0.6167	0.5649	0.0518	Ecuador	0.4869	0.5176	−0.0307
Brunei Darussalam	0.6250	0.7699	−0.1449	Egypt	0.4611	0.4728	−0.0117
Bulgaria	0.6132	0.5850	0.0282	El Salvador	0.5513	0.4916	0.0597
Equatorial Guinea	0.2955	0.7239	−0.4284	Ireland	0.7149	0.7360	−0.0211
Eritrea	0.2043	0.1652	0.0391	Israel	0.8100	0.6967	0.1133
Estonia	0.7987	0.6416	0.1571	Italy	0.7190	0.6969	0.0221
Ethiopia	0.2306	0.2250	0.0056	Jamaica	0.4552	0.5252	−0.0700

(continued)

Table 7.1 (continued)

Country	EGDI	EEGDI	EGD-GDP	Country	EGDI	EEGDI	EGD-GDP
Fiji	0.4672	0.4323	0.0349	Japan	0.8019	0.7161	0.0858
Finland	0.8505	0.7254	0.1251	Jordan	0.4884	0.4624	0.0260
France	0.8635	0.7141	0.1494	Kazakhstan	0.6844	0.5749	0.1095
Gabon	0.3687	0.6049	-0.2362	Kenya	0.4212	0.2968	0.1244
Gambia, The	0.2688	0.3169	-0.0481	Kiribati	0.2998	0.4759	-0.1761
Georgia	0.5563	0.4510	0.1053	Korea, South	0.9283	0.7059	0.2224
Germany	0.8079	0.7295	0.0784	Kuwait	0.5960	0.7788	-0.1828
Ghana	0.3159	0.3701	-0.0542	Kyrgyzstan	0.4879	0.3341	0.1538
Greece	0.6872	0.6899	-0.0027	Laos	0.2935	0.3531	-0.0596
Grenada	0.5479	0.5719	-0.0240	Latvia	0.6604	0.6140	0.0464
Guatemala	0.4390	0.4401	-0.0011	Lebanon	0.5139	0.5920	-0.0781
Guinea-Bissau	0.1945	0.2412	-0.0467	Lesotho	0.3501	0.2826	0.0675
Guyana	0.4549	0.4963	-0.0414	Liberia	0.2407	0.1185	0.1222
Haiti	0.1512	0.2381	-0.0869	Lithuania	0.7333	0.6326	0.1007
Honduras	0.4341	0.4188	0.0153	Luxembourg	0.8014	0.8465	-0.0451
Hungary	0.7201	0.6357	0.0844	Macedonia	0.5587	0.5440	0.0147
Iceland	0.7835	0.7336	0.0499	Madagascar	0.3054	0.2075	0.0979
India	0.3829	0.3957	-0.0128	Malawi	0.2740	0.2156	0.0584
Indonesia	0.4949	0.4310	0.0639	Malaysia	0.6703	0.6044	0.0659
Iran	0.4876	0.5674	-0.0798	Maldives	0.4994	0.5174	-0.0180
Iraq	0.3409	0.4005	-0.0596	Mali	0.1857	0.2415	-0.0558
Malta	0.7131	0.6766	0.0365	Poland	0.6441	0.6391	0.0050
Mauritania	0.1996	0.3184	-0.1188	Portugal	0.7165	0.6609	0.0556
Mauritius	0.5066	0.5980	-0.0914	Qatar	0.6405	0.8791	-0.2386
Mexico	0.6240	0.5947	0.0293	Romania	0.6060	0.5858	0.0202
Moldova	0.5626	0.3830	0.1796	Russia	0.7345	0.6116	0.1229
Mongolia	0.5443	0.4274	0.1169	Rwanda	0.3291	0.2407	0.0884
Montenegro	0.6218	0.5567	0.0651	Samoa	0.4358	0.4648	-0.0290
Morocco	0.4209	0.4398	-0.0189	São Tomé and Príncipe	0.3327	0.3063	0.0264

Country			
Mozambique	0.2786	0.2133	0.0653
Myanmar	0.2703	0.2854	−0.0151
Namibia	0.3937	0.4811	−0.0874
Nepal	0.2664	0.2584	0.0080
Netherlands	0.9125	0.7456	0.1669
New Zealand	0.8381	0.6841	0.1540
Nicaragua	0.3621	0.3762	−0.0141
Niger	0.1119	0.1643	−0.0524
Nigeria	0.2676	0.3428	−0.0752
Norway	0.8593	0.7782	0.0811
Oman	0.5944	0.6926	−0.0982
Pakistan	0.2823	0.3525	−0.0702
Panama	0.5733	0.5904	−0.0171
Papua New Guinea	0.2147	0.3325	−0.1178
Paraguay	0.4802	0.4584	0.0218
Peru	0.5230	0.5397	−0.0167
Philippines	0.5130	0.4138	0.0992
Sweden	0.8599	0.7380	0.1219
Switzerland	0.8134	0.7492	0.0642
Syria	0.3705	0.4397	−0.0692
Tajikistan	0.4069	0.3148	0.0921
Tanzania	0.3311	0.2634	0.0677
Thailand	0.5093	0.5306	−0.0213
Timor-Leste	0.2365	0.3726	−0.1361
Togo	0.2143	0.2228	−0.0085
Tonga	0.4405	0.4989	−0.0584
Trinidad and Tobago	0.5731	0.6410	−0.0679
Tunisia	0.4833	0.5319	−0.0486
Turkey	0.5281	0.5882	−0.0601
Turkmenistan	0.3813	0.5091	−0.1278
Tuvalu	0.3539	0.3959	−0.0420
Saudi Arabia	0.6658	0.6701	−0.0043
Senegal	0.2673	0.3062	−0.0389
Serbia	0.6312	0.5536	0.0776
Seychelles	0.5192	0.6661	−0.1469
Sierra Leone	0.1557	0.1843	−0.0286
Singapore	0.8474	0.7978	0.0496
Slovak Republic	0.6292	0.6604	−0.0312
Slovenia	0.7492	0.6900	0.0592
Solomon Islands	0.2416	0.3787	−0.1371
South Africa	0.4869	0.5521	−0.0652
Spain	0.7770	0.6993	0.0777
Sri Lanka	0.4357	0.4549	−0.0192
St. Kitts and Nevis	0.5272	0.6215	−0.0943
St. Lucia	0.5122	0.5759	−0.0637
St. Vincent and the Grenadines	0.5177	0.5705	−0.0528
Suriname	0.4344	0.5326	−0.0982
Swaziland	0.3179	0.4409	−0.1230
Uganda	0.3185	0.2559	0.0626
Ukraine	0.5653	0.4912	0.0741
United Arab Emirates	0.7344	0.7040	0.0304
United Kingdom	0.8960	0.7233	0.1727
United States	0.8687	0.7643	0.1044
Uruguay	0.6315	0.6086	0.0229
Uzbekistan	0.5099	0.3822	0.1277
Vanuatu	0.3512	0.4439	−0.0927
Venezuela	0.5585	0.5829	−0.0244
Vietnam	0.5217	0.3843	0.1374
Yemen	0.2472	0.3484	−0.1012
Zambia	0.2910	0.2825	0.0085
Zimbabwe	0.3583	0.0981	0.2602

Indeed, the UAE's per capita GDP has increased markedly over the same time period so one would expect the EGDI to improve accordingly, yet when per capita GDP is taken into account the EGDGDP value fell slightly from 0.0369 in 2008 to 0.0304 in 2012. In other words, although the UAE has a higher level of e-government development than would otherwise be expected given the country's GDP, the extent to which this is the case actually fell between 2008 and 2012. While there are other countries whose performance relative to that which would otherwise be expected increased over the same time-frame, it is the UAE that is noted in the UN report. This is not denigrate the improvement that has been achieved in the UAE – it should not be forgotten that the EGDI score is still positive – rather to note that there are other countries whose performance is even better and would be more suitable examples of 'best practice'.

Similarly, the UN report considers the Seychelles to be a leader in Eastern Africa, yet after per capita GDP is taken into account the Seychelles is an underperformer with an EGDGDP $= -0.1469$. On the other hand, another country in Eastern Africa – Zimbabwe – has the world's highest EGDGDP but their performance relative to GDP goes without mention in the report.

Academic sources also often cite countries' EGDI scores as evidence of those countries' performance, or lack thereof, in electronic government. For example, Bhuiyan (2010) notes the low overall ranking given to Bangladesh, yet when one considers the trajectory of Bangladesh's e-government development relative to GDP the situation is positive: EGDGDP rose from -0.2136 in 2008 to -0.0076 in 2012, one of the highest increases in performance globally.

Similarly, Al-Wazir and Zheng (2012) imply that e-government performance in Yemen is poor and that Yemen could learn from other countries such as Egypt. Yet Yemen's performance relative to GDP increased by more than Egypt's from 2008 to 2012 – Yemen's EGDGDP increased by 0.1288, in contrast to Egypt's 0.0479. The implication is that the strategies being employed in Yemen have led to a more rapid increase its performance relative to GDP, and by extension it can be argued that e- policies to improve e-government development in Yemen are more effective than those in Egypt.

There are many other examples where the UN rankings are reported without question are also present in academic literature (e.g. Assar et al. 2011; Opesade 2011). Given the examples discussed above, academic debate that does not take GDP into account may be unreliable and misleading. Further, information from non-academic sources in the public domain also use the raw EGDI data; for example, the Wikipedia entry on e-government includes a list of the top 50 countries based on the UN rankings.

Many national governments have also repeated the UN rankings where it suits them, for example government websites from Mauritius,[1] Singapore,[2] Saudi Arabia[3]

---

[1] http://www.gov.mu/portal/sites/indicators/International_Indices.html

[2] http://www.egov.gov.sg/accolades-and-awards-international-awards

[3] http://www.alriyadh.gov.sa/en/news/Pages/news8668.aspx

and Qatar[4] all cite the EGDI as flattering evidence of the state of e-government in those countries, yet when GDP is taken into account they fared worse than implied by their websites, and in some cases much worse.

Clearly, although in many cases the raw UN rankings may give a misleading indication of a country's performance in e-government development, these rankings are a staple of discussion in official, academic and public discourse. This creates a risk that the wrong exemplars will be put forward in policy and related debates. Given the clear correlation with a nation's level of wealth and its level of e-government development, there is no reason that a more accurate ranking that takes this into account cannot and should not be devised given the simplicity of doing so.

This section has incorporated GDP into EGDI to better assess countries' E-government Development Performance, resulting in a metric termed EGDGDP. However, in addition to economic factors, sociocultural factors are also an important aspect that must be considered when considering a country's e-government development. In the following section these sociocultural factors will be explored in more detail.

7.3   Sociocultural Factors

Over the past two decades or so many countries around the world started to utilize the power of ICT to deliver public services. Despite the obvious benefits of online services; the uptake of these services is still limited in many countries (Van Deursen et al. 2006; Kunstelj 2007; Bertot and Jaeger 2006; Ebbers et al. 2008). Further, user participation in online services is a key success factor in enabling e-government to reach its full potential (Moon and Norris 2005; Jaeger 2003).

The limited e-government user uptake has been attributed to two main factors. First, e-government activities have been predominantly driven by supply side factors (Bertot and Jaeger 2006; Kunstelj et al. 2007; Reddick 2005; Schedler and Summermatter 2007; Gareis et al. 2004; Ebbers et al. 2008). Many governments' decision to offer e-services was predominantly influenced by supply side factors such as cost and time cutting (Anthopoulos Siozos and Tsoukalas 2007). During the early e-government initiatives it seems that a number of governments decided to deliver their traditional offline services online. ICT was mainly used to automate existing services following the same existing business process (Asgarkhani 2005) rather than using ICT to radically redesign their services. Governments failed to use the power of ICT to invent new possibilities.

Second, e-government initiatives are technology driven rather than needs driven (Bertot and Jaeger 2008). Governments are motivated by what new technology can

---

[4] http://portal.www.gov.qa/wps/wcm/connect/hukoomi+web+content/hukoomi/media+center/news+and+press+releases/individual+news/hassan+al-sayed+it+increases+government+productivity

offer to deliver public services rather than actual users' needs and requirements. Verdegem and Verleye (2009) among others call for a more comprehensive understanding of users' needs and satisfaction to improve users' uptake of e-services.

The OECD (2009) report stated that governments need to improve the uptake of e-services by transforming their e-government strategies. The report recognizes the shift of focus and approach of online public services from government-centric to user-centric by placing more attention to the context, such as social, organization, and institutional factors, in which e-government services are delivered is needed.

This implies that the sociocultural factors in which e-government projects are implemented must be considered from both the supply side (governments) and the demand side (users).

7.3.1 Sociocultural Factors and the UN Report

The UN report described EGDI as being a measure of the 'willingness and capacity' of national governments to use online methods to provide government services (UN 2012). While it is not clear exactly what 'willingness' means at a national level, it is likely that national culture might be a relevant factor. Zhao (2011) examined the relationship between national culture and EGDI and found that smaller power distances, higher levels of individualism and higher long-term orientation are all related to increased e-government development; it is possible that these might have some relationship with the willingness that EGDI attempts to measure.

However, it may be a mistake to make conclusions about countries' willingness. Countries with lower per capita GDP might be extremely willing to increase their e-government readiness but lack capacity due to economic constraints.

While the UN report does not explicitly talk about the impact of countries culture on their e-government development, the report recommends government departments and agencies to change their culture in the form of promoting information sharing, cooperative knowledge management to make the transformation to a more citizen-centric approach possible. The report does not address the cultural changes required towards e-government from the users' side, including citizen trust in government and their willingness to voluntary participate in online services. It is arguable that such a change is critical factor in shaping users' attitudes towards governments' ITeS because as Evans and Yen (2006) note, without a deep unwavering trust of citizens in their government there will not be a free flow of information.

The UN (2012) report states that governments have focused predominantly on the provision of online services from a supplier perspective. However, the report also acknowledges that recently there has been a shift towards a more consumer demand driven policy and greater emphasis on citizen usage of online services. Despite this recent shift, it has been stated that the uptake of online services is still low. This slow uptake has been attributed to accessibility of online services issues with only 24 countries openly promote free access to e-government services through free Wi-Fi or kiosks. The report also suggests that the lack of using social media to leverage the adoption of e-services explains the low uptake. The report states that only 40 % of governments use a social networking site.

The UN report explores the role sociocultural factors play in influencing the uptake of e-government services by measuring each country's Human Resource Capital index. The Index is a weighted average that is composite of two indicators: the adult literacy rate and the combined primary, secondary, and tertiary gross enrolment ratio, with two thirds of the weight assigned to adult literacy rate and one third of the weight assigned to the gross enrolment ratio. However, a number of sociocultural factors such as power distances, individualism, government transparency, citizen's trust in government, and citizen's privacy concerns are not included.

We argue that there is a need to adjust the Human Capital index by introducing not only the adult literacy rate and the enrollment ratio (which could be either demand or supply side factor but mainly focus on the supply side) but also adding other sociocultural perspectives such as privacy and trust to achieve better e-government development indicators.

The UN 2012 acknowledges the privacy and security challenges facing many countries. The report states that the majority – 59 % – of governments' websites around the world, mostly from developing countries, lack a privacy policy. Yet the existence or absence of this feature is not part of the EGDI. The report also states that almost half of the countries in Europe display secure links on their national websites, while only one in Africa appears to do so. However, the existence or absence of such security features in the website is also not part of the EGDI.

Interview data from a current study conducted by one of the authors that investigates factors that influence individuals' uptake of e-government in Abu Dhabi, United Arab Emirates (UAE) suggest that privacy concerns, trust in government and security concerns are all influencing factors that shape potential e-government user decisions to use or not to use online services. The results clearly indicate that end-users who perceive government to be trustworthy and transparent are likely to trust and use the government's online services.

Yet these demand side factors are overlooked by the 2012 UN report, which suggests that offering more online public services will create more demand for the services by end users:

> E-government innovation and development can position the public sector as a driver of demand for ICT infrastructure and applications in the broader economy (p. 10).

Promoting the uptake of online services has been considered only from the supply side. We acknowledge the provision of ICT infrastructure and e-government application are important factors in e-government development around the world, but improving citizens' trust in their government in the first place is of equal importance as people who hold positive views about their government are more likely to attempt to use online services.

Cultural background and communication preference in addition to the Human Capital components play a role in their acceptance and use of online services. Results from the current study noted above showed that participants from the

Middle East and the Far East tended to prefer to communicate with government in face-to-face interactions, while participants from a western backgrounds often preferred to use online services and believed it conferred relative advantages such as convenience, saving time, and so on. The point here is that e-government developments must be considered not only by measuring the EGDI but also taking into account individual member state culture; the context in which e-government is implemented varies from one county to another.

7.3.2   Implications of Considering Sociocultural Factors

It is clear from the above discussion that the sociocultural aspects of different countries are not considered by the UN report in assessing e-government development in countries around the world. The EGDI is not telling the full story; therefore the index needs to be adjusted to take into account cultural and economic factors. While the index has been renamed 'development', it still measures e-government readiness. Including these factors in the index will allow the creation of a better index of a country's e-government readiness that takes into account not only the supply of e-government services but also the demand and acceptance for such services.

When monitoring countries' e-government developments, relying only on the supply focused EGDI, and overlooking the sociocultural aspects could lead to false sense of advancement in some countries. The lack of consideration to sociocultural factors when considering and make decisions about government online services could lead to the conceptions and implementation of e-government systems that are either not used or underutilized by end-users. The focus should always be on the users' needs and expectations as users are the most important tools in implementing ITeS.

7.4   Conclusions

If a country is not well developed there may not be services which are amenable to online delivery in the first place. For example, if a country does not have a 'Medicare' style program there is little possibility of providing an e-government interface to such a program.

Hence, the UN data need to be interpreted carefully before any specific recommendations or policy decisions are made. The level of public discourse and debate about e-government seems prone to considering e-government development in absolute values and not relative to each individual country's economic and sociocultural circumstances.

Each individual country needs to look not only at the level of development of their telecommunications infrastructure, the literacy and technological skills available in that country, and the e-government services already in place, but should also look at the ability the country has to afford the development of new e-government services and the sociocultural appropriateness of any proposed services. To fail to do this increases the risk that inappropriate investment and policy decisions are made.

There is also a need for greater attention to be paid by academic and industry researchers to the causes for countries' differing levels of performance in e-government development. Wealth and sociocultural factors clearly play a role, but further investigation into how these factors influence e-government development is necessary. Without it, the benefits promised by e-government will be more difficult to realise, and perhaps not achieved at all.

References

Al-Wazir AA, Zheng Z (2012) E-government development in Yemen: assessment and solutions. J Emerg Trends Comput Inf Sci 3(4):512–518

Anthopoulos LG, Siozos P, Tsoukalas IA (2007) Applying participatory design and collaboration in digital public services for discovering and re-designing egovernment services. Gov Inf Q 24:353–376

Asgarkhani M (2005) The effectiveness of e-service in local government: a case study. Electron J e-Gov 3(4):157–166

Assar S, Boughzala I, Boydens I (2011) Back to practice: a decade of research in e-government. In: Assar S et al (eds) Practical studies in e-government: best practices from around the world. Springer, New York, pp 1–12

Bertot JC, Jaeger PT (2006) User-centered egovernment: challenges and benefits for government web sites. Gov Inf Q 23(2):163–168

Bertot JC, Jaeger PT (2008) The e-government paradox: better customer service doesn't necessarily cost less. Gov Inf Q 25(2):149–154

Bhuiyan MSH (2010) E-government applications in Bangladesh: status and challenges. In: Davies J, Janowski T (eds) Proceedings of the 4th international conference on theory and practice of electronic governance (ICEGOV '10). ACM, New York, pp 255–260, doi 10.1145/1930321.1930374, http://doi.acm.org/10.1145/1930321.1930374

Ebbers WE, Pieterson WJ, Noordman HN (2008) Electronic government: rethinking channel management strategies. Gov Inform Quart 25(2):181–201

Evans D, Yen D (2006) E-government: evolving relationship of citizens and government, domestic, and international development. Gov Inf Q 23(3):207–235

Gareis K, Cullen K, Korte W (2004) Putting the user at the center. Implications for the provision of online public services. In: Cunningham P (ed) eAdoption and the knowledge economy. Issues, applications, case studies. IOS Press, Amsterdam, pp 611–618

Jaeger PT (2003) The endless wire: e-government as global phenomenon. Gov Inform Quart 20:323–331

Jaeger PT, Thompson KM (2003) E-government around the world: lessons, challenges, and future directions. Gov Inform Quart 20:389–394

Kunstelj M, Jukic T, Vintar M (2007) Analysing the demand side of E-government: what can we learn from Slovenian users? vol 4656. Springer, Berlin/Heidelberg, pp 305–317

Moon MJ, Norris D (2005) Does managerial orientation matter? The adoption of reinventing government and e-government at the municipal level. Inform Syst J 15:43–60

OECD (2009) Rethinking e-government services: user-centred approaches. OECD, Paris

Opesade AO (2011) Strategic, value-based ICT investment as a key factor in bridging the digital divide. Inform Dev 27(2):100–108

Reddick CG (2004) A two-state model of E-government growth: theories and empirical evidence for US cities. Gov Inform Quart 21:51–64

Reddick CG (2005) Citizen interaction with e-government: from the streets to servers? Gov Inform Quart 22(1):38–57

Schedler K, Summermatter L (2007) Customer orientation in electronic government: motives and effects. Gov Inform Quart 24(2):291–311

Traunmüller R, Wimmer MA (2003) E-government at a decisive moment: sketching a roadmap to excellence. Lect Notes Comput Sci 2739:1–14

Uesugi S (2008) Bridging between real and virtual – technologies to advance ITeS. In: Proceedings of the international symposium on applications and the internet. Turku, 28 July–1 Aug, pp 440–443

United Nations (2012) United Nations e-government survey 2012: e-government for the people. United Nations, New York

van Deursen A, van Dijk J, Ebbers W (2006) Why e-government usage lags behind: explaining the gap between potential and actual usage of electronic public services in the Netherlands. Lect Notes Comput Sci 4084:269–280

Verdegem P, Verleye G (2009) User-centered e-government in practice: a comprehensive model for measuring user satisfaction. Gov Inf Q 26(3):487–497

Weerakkody V, Janssen M, Dwivedi YK (2009) Handbook of research on ICT-enabled transformational government: a global perspective. Information Science Reference, New York

West DM (2004) E-government and the transformation of service delivery and citizen attitudes. Public Adm Rev 64(1):15–27

Whitmore A (2012) A statistical analysis of the construction of the United Nations e-government development index. Gov Inform Quart 29:68–75

Zhao F (2011) Impact of national culture on e-government development: a global study. Internet Res 21(3):362–380

Chapter 8
Computer Mediated Communication and Telecollaboration for Language Learning: Issues of Technology

Paul Spijkerbosch

8.1 Preamble

Education is a critical element of most contemporary societies. The level, quality and type of education that individuals receive can have a major influence on both the types of job positions they may acquire and the earnings that they may generate. Furthermore, lifelong learning is important in acquiring new knowledge and upgrading one's skills, particularly in this age of rapid technological and economic changes. The educational services industry includes a variety of institutions that offer academic education, vocational or career and technical instruction, and other education and training to millions of students each year (Wilhelm 2011). And technological proficiency, most would argue, is critical.

Yet technology is not just about innovations; it is about service, access, literacy, and relevance. If these are missing or deficient in any way, adoption is likely to be constrained. This construct is certainly relevant in the education industry. And it stretches across borders. No longer, for example, are gadgets being priced cheaply only for learners in developing countries. The Raspberry PI, priced below 50 American dollars and developed by a charity for users in the United Kingdom, is hardware designed to provide Internet connectivity and app developmental capability at low cost for children at a fraction of the cost of smartphones currently out on the market. This nexus of access, literacy, and relevance has ensured the Raspberry PI being sold out, with further production runs being planned at an even lower user cost (Davison 2012).

When people think of technology, invariably they think of computers, machines and other similar artifacts that either simplify or enable their day-to-day life. In an intriguing reconceptualization of technology, Zhouying Jin (2011) asserts that people need to have a much wider appreciation of what technology means to

P. Spijkerbosch (✉)
Department of Sociology, Faculty of Humanities, Matsuyama University, Matsuyama, Japan
e-mail: spijker@cc.matsuyama-u.ac.jp

S. Uesugi (ed.), *IT Enabled Services*,
DOI 10.1007/978-3-7091-1425-4_8, © Springer-Verlag Wien 2013

people. Drawing on philosophers such as Plato, to research institutes such as Nomura, she elegantly argues that people need to widen their concept of technology to include techniques, approaches, and processes (Zhouying 2011). She describes this as 'soft tech'. Although her perspective is clearly not unique, what is indisputable is that although artifacts have enabled changes, people need to develop ways to best make these changes work. It is this factor that is driving a lot of research in social sciences of late. Zhouying Jin believes that these changes can be described as traditional notions of 'hard tech' being supplanted by wider notions of 'soft tech'.

This paper will outline how important this 'soft tech' is in appropriately incorporating 'hard tech' innovations within the educational service industry. In fact, it will quickly become apparent just how complicated 'soft tech' solutions may be, compared to that of 'hard tech'. Specifically, it will look at how Internet telephony can be best managed for intercultural collaboration and language learning.

8.2   Background

Rapid and on-going developments in computer-mediated communication (CMC) technologies increasingly facilitate opportunities for language learners, educators, and educational institutions. Unsurprisingly, these opportunities have been taken advantage of in a variety of ways. Initially they were (and still are) used for e-learning or communicative forums through emails, websites or blogs. Recently, widespread access to Web 2.0 technologies such as VoIP telephony (Skype, Google Talk, Messenger) or 3D avatar software (Second Life, Active Worlds) has become available.

Language learning has traditionally been described in terms of location: foreign language (FL), whereby the learning is undertaken in a country where the language is not commonly used, or second language (SL) acquisition, where the learning takes place in a country in which the target language is commonly used in day-to-day transactions. Naturally SL environments provide opportunities for learners to absorb a more holistic range of target (such as paralinguistic or pragmatic) skills. In contrast, FL teachers cannot easily recreate a SL style learning environment: it is usually bound to the classroom, creating debate over the authenticity of the learning experience.

The reason CMC technologies have become so meaningful for language acquisition is because they can facilitate communication in a number of forms which together, can provide learners with more of the opportunities previously ascribed to the SL environment, despite being accessed in the FL environment.

Although CMC technologies can enable more 'authentic' language learning opportunities, their primary practical use seems mostly defined by transactional activities: getting learners to negotiate meaning through interaction. Accordingly a lot of research has focused on Telecollaboration, in which collaborative exercises are undertaken via internet telephony. As often happens however, coined phrases can lead to a range of meanings. For the purposes of this research, Telecollaboration will be defined as a project in which learners need to (with differing levels of scaffolding) work with Internet-based partners to reach a common goal. The language component may be dependent on learner proficiency and whether the project involves a single language or language exchange (commonly referred to as e-Tandem).

Initially, approaches to CMC technologies have been primarily driven by individual instructors at institutions. Recently, as CMC financial and technological barriers become lower, educational institutions seem to be joining the bandwagon. Ensuring practical and effective strategies for incorporating these technologies are put in place, however, remains one of the biggest concerns for stakeholders. Although there is research detailing types of collaboration and their assessment, little seems to touch on the need for curricula or blended subjects (combining an academic subject with the target language).

8.3 Barriers

It is easy to be caught up in the novelty of using technological innovations to communicate with others. Ever changing fads and quirks mushroom in the information technology sector, with various conferences and expositions given the kind of attention previously afforded to Cannes and Hollywood. Educators are no different, and with good reason. Technological innovations offer a variety of methods to facilitate language learning. But, as Warschauer (1996) warned, technology itself does not improve language learning, but rather, it is the manner in which it is utilized. There are a number of technologically related issues that would need to be ameliorated for a truly successful CMC project to eventuate.

8.3.1 Multimodal Literacy

Just like literacy defines ability to read and write language, multimodal literacy signifies the ability to understand and utilize a range of technologically-modified communicative modes appropriately.

Despite technology fast becoming an integral part of most people's lives, exposure can elicit range of abilities, and not all users are proficient; or rather, proficient at some, but inadequate at others. Furthermore, as the penetration of technology in our society continues, the emergence of multi-modal skills in CMC use is becoming increasingly important, not only for the learner, but also for the instructor in the classroom.

Multimodal literacy, defined by Pegrum (2009) as 'understanding and interpreting the relationship and interaction between different formats of digital media', is both gateway and barrier to language learning – in the classroom at least (Guth and Helm 2010). In a study of task-based language teaching (TBLT) course design reflection,

Hauck (2010) explored the interrelationship between multimodal literacy and online communication and concluded that educators need to be teachers of the technology, not just facilitators, if their learners are going to be able to fully partake of the opportunities for language learning and intercultural development (Hauck 2010). In other words, if teachers are going to expect their students to use technological applications (viz. CMC) it would be appropriate to ensure that all students can utilize them effectively. Common sense maybe, but nevertheless literature suggests that it remains a significant hurdle.

8.3.2 Appropriacy

In a world that seems to rapidly churn out all manners of technological marvels and innovations, teachers need to be able to cherry pick the most relevant options for their classroom: options that enhance, rather than distract from, the learning process for their students.

In choosing a collaborative technology, instructors should determine how much, and what type, of student interaction is needed to complete group assignments and facilitate learning (Parker and Ingram 2011). As referred to previously, functionality can be exciting, but it is only one part of the selection criteria. Timmerman and Kruepke (2006) point out that more features are not necessarily better. Function availability doesn't equate to student usage. Having too many tools – or tools with a steep learning curve – can impede, rather than facilitate, student learning (Falowo 2007). Otherwise, as Loveless et al. (2001) point out, effective learning through integrated use of Internet Communicative Technologies (ICT) is likely to occur despite, and not because of, the role of the teacher. One example of this distractive quality is the recent usage of avatar-based CMC (such as in Second Life or Active Worlds), after which some students concluded that they had been sidetracked by the novelty and that they might have gotten better return by just sticking to simplified chat forums (Deutschmann et al. 2009).

8.3.3 Accessibility

Despite the recent ubiquity of CMC, accessibility remains an issue. Accessibility hinges on a number of aspects, such as; age, location, and time.

Learners are of all ages, and accordingly CMC may be utilized at all age levels. However, younger learners often need more teacher-centered pedagogy to participate productively in class, while older participants may feel marginalized with regard to technology. What this means is that age may affect how accessible the CMC is perceived by learners. Perception can be a powerful psychosomatic realization despite environmental factors that may indicate otherwise.

Parker and Ingram (2011) point out that there are a variety of micro or macro issues that may affect participatory rates. Classroom ambience and social dynamics can affect the development of class community – an important factor for teachers to take into account considering their focus on collaborative tasks.

> ...if technology is truly experienced differently by different users, then the effects may vary by user as well, and studying its effects at multiple levels of analysis is a necessity. [One option] may be to examine how instructors can best move students beyond learning the chosen technologies to learning how to use their functions to collaborate effectively, no matter what technology is used. Parker and Ingram (2011)

Some telecollaboration involves different time zones. Students in a Japanese secondary school have little chance to communicate live with peers, for example at a North American school, forcing them to fall back on delayed methods of telecollaboration. Class scheduling can exacerbate this issue.

Many students' access to CMC is restricted outside the classroom for a variety of reasons: bandwidth, hardware, parental concerns. Consequently, accessibility through either bandwidth or portal outside the classroom can often be problematic.

8.4   Philosophical Frameworks

Rod Ellis in a foreword to Thomas and Reinders (Ellis 2010) elegantly details what he describes as interactionist theories that underpin the raison d'etre of SLA within CMC environments. He considers that most research to date in this field has been informed by 'negotiation-of-meaning sequences that support learning by providing comprehensible input, feedback and opportunities for learners to self-correct' (Ellis 2010). He goes on to point out however, that learners using CMC have communicated in different ways from traditional classrooms, requiring researchers to understand and describe why this is so. Although the research to date may well be grouped in terms of 'interactionist theories', there are big differences within this purview.

Research into international exchanges with CMC was initially framed in terms of cognitive approaches. More recently, this has been superseded by socio-cultural frameworks (Lantolf and Thorne 2006; Lamy and Hampel 2007). A third approach has been to focus on tasks within the CMC moderated exchange.

The cognitive approach, championed by Chomsky's assertion that mind and matter were separated, considered information technology to epitomize the move of thought and rationalization from meatspace to cyberspace. Language represented rationalization, and thus, CMC would empower people from different cultures to minimize their socio-cultural restraints and enable them to improve their language learning without socio-cultural 'hindrance'. These theories quickly dissipated from the realization users' actions, and apparent thoughts, could not be divided so easily into the dichotomy of mind and body. Rather their thoughts seemed to reflect their environment or past influences. This quickly led to the development of frameworks that could better explain these phenomena.

Socio-cultural theories (SCT) stress social interaction for learning. Social interaction, through participation in cultural, linguistic, and historically formed settings such as family life and peer group interaction, and in institutional contexts like schooling, organized sports activities, and work places, leads to development of language – as language is a fundamental aspect of interaction (Lantolf and Thorne 2006). Key aspects of SCT involve mediation, regulation, and abstraction (Lantolf and Thorne 2006), based on the principles of SCT set out by Vygotsky.

Another important aspect of SCT within the context of SLA is the zone of proximal development (ZPD) (Donato 2000; Ohta 2001). The ZPD is defined as the zone of potential in which an individual can achieve more with assistance from others with better proficiency than they can do alone. The significance of this notion is that learning is linked to development only within the ZPD. Facilitating contact between language users of different abilities to help create this ZPD is a principal goal of educators using CMC technologies (Cheon 2008).

The nature of CMC has invoked further rationalization of the processes involved in SCT: multimodalities and multiliteracies. Lamy and Hampel (2007), quoting Wertsch (2002) and Kress (2003), point out that the functionality of CMC ensure that modes of communication transform previous modes to such an extent that they may be totally different with respect to the affordances they represent to users (Smith 2003, 2005). This raises the issue of whether or not users are able to adjust accordingly. The plethora of modes is only matched by the need for skills, or literacies, to use them appropriately. In other words language requires not only certain levels of competency, but also various other skills – such as technological or communicative competence. Deficiencies in one or more of these additional skillsets can negatively impact on the likelihood for successful language learning within a CMC environment.

A modern derivation of Vygotsky's work (Müller-Hartmann and Schocker-v. Ditfurth 2010) that incorporates these further rationalizations has become popularized as Activity Theory (AT). AT encompasses multimodalities and multiliteracies within a number of facets and levels. Importantly, it facilitates a comprehensive research framework for the pedagogical implications of Task-Based Language Learning (TBLL) in the CMC classroom.

As SCT has dominated the approaches of recent research frameworks, cultural interaction as a trigger for developing intercultural competence (IC) seems to have become an end in itself, rather than a means, for a lot of SLA researchers. Researchers have drawn on the behaviorist psychology concepts of incidental and intentional learning, popularized during the 1970s, to develop tools that could describe the language learning processes within CMC environments (Kabata and Edasawa 2011; Hulstijn 2003). Kabata and Edasawa (2011) quotes Huckin and Coady (1999) in arguing that "incidental acquisition is the primary means by which second language learners develop their vocabulary beyond the first few thousand most-common words". In fact, incidental learning is believed to have certain advantages over direct instruction (Kabata and Edasawa 2011) as it is contextualized, learner centered, and is pedagogically efficient (vocabulary acquisition and reading occur at the same time).

8.5   Management

8.5.1   Blending Subjects

Subject division was a mechanism developed by philosophers in order to clearly define bodies of knowledge. Increasingly however, these boundaries are being blurred as topics of interest draw on disparate subjects to transmit knowledge. Points of convergence between these subjects have automatically led to ever growing fragmentation and specialization, evidenced by the growing number of 'subjects' available for study at all stages of formal education. This process has been accelerated through information technology as knowledge becomes more accessible. Although widely recognized by educators, the retention of these constructs is considered a 'necessary evil' however, as they provide a form of control.

To reconceptualize this within a Japanese context, English language education is typically taught in complete isolation to other subjects in the school system. English teachers rarely, if at all, teach in conjunction with instructors of other subjects. This is not to say English, or language instruction, is the exception. Most subjects are taught in isolation. Although this may be appropriate in some situations, educators and researchers need to consider if recognizing these points of convergence in a curriculum would be more beneficial for learners. Certainly with regard to CMC telecollaboration, points of convergence would allow more instructor interaction at the curricular level, and enable time saved on avoiding repetition to be spent on other worthwhile topics.

8.5.2   Collaboration

It could be argued that a number of factors are leading many teachers to narrow their focus at a time when they should be scaffolding their students and setting an example in the classroom by encouraging collaborative practices. Factors that complement the narrowing of focus include more emphasis on performance measures (Gewirtz 2002), teaching to the test, and skewing students' integrated knowledge about language (Frater 2000), and limited time to work with others outside their specialized area of expertise (Hodkinson and Hodkinson 2005). With specific regard to collaborative practices, Gereluk (2005) notes that:

> Collaboration requires time and effort amongst staff and a demanding curricular framework may overwhelm an already overworked teacher. The inflexibility of the curriculum may create a situation whereby teachers do not have time to collaborate or see the need to collaborate when every detail has been laid out.

8.5.3 Curricular Development

Curriculum has come to represent different concepts to different people. Whether it is transmitted, a product, a process or praxis, curricula seldom makes all stakeholders happy. Furthermore, it could be argued, curricular theory can distract teachers from the art of teaching: learning too often occurs in spite of, rather than because of, instruction. Nevertheless, for learning environments typically provided by institutions, a curriculum remains the framework through which potential value can be appraised by stakeholders.

Around the world, curricula have become a mechanism for the transmission of social values. This then, can be seen as a cultural construction. Teachers attempting to establish a telecollaborative venture need to ameliorate their different educational (or cultural) systems to facilitate common goals.

8.5.4 Competencies

Hauck (2010) outlines what she describes as the 'interdependence of multimodal and intercultural communicative competencies'. Using Internet-based telephony to collaborate interculturally requires intercultural communicative skills as well as technological skills. They are dependent upon each other, and checking and scaffolding learner knowledge of them needs to be considered fundamental if they are to be effectively used pedagogically.

8.5.5 Intercultural Communicative Competence (ICC)

A number of researchers in various fields have addressed the need for ICC. However, with regard to language learning in conjunction with ICC, Byram (1997) developed the seminal model. Specifically, Byram (1997) considered language use to be a manifestation of culture (Thorne and Lantolf 2007). In other words, he argued that language could not be separated from culture. To describe this in more detail, Byram outlined five types (or savoirs) of competencies that language users employ in variable quantities at different times: Attitudes, Knowledge, Discovery and interaction, Interpreting and relating, and Critical cultural awareness – all of which are constructs that have been well defined. Furthermore, he outlined more than 20 specific classroom objectives, most of which are performance-based. These objectives have ensured that Byram's model is particularly useful for institutions (which tend to be objective-based). Furthermore the model is grounded in student-centered literature, and focuses on concepts of language mastery.

Conversely, Deardorff's (2006) study, based on interviews with a group of educational administrators and a group of ICC 'experts', found that although specific ICC objectives were favored by the administrators, the experts leant toward a more generalist set of benchmarks. Nevertheless, the experts in Deardorff's study were able to reach a consensus on certain attributes that could define ICC; attributes that focused on cognitive and interactional dispositions. They could not reach consensus on what role language had to play in the acquisition or performance of ICC however, although it was recognized as *being critical* (my own italics). Regardless, what was significant in terms of cross study reliability is that Deardorff reached similar conclusions in comparison to a study by Fantini (2006) in which Fantini noted that "a complex of abilities [is] needed to perform effectively and appropriately when interacting with others who are linguistically and culturally different from oneself ".

8.5.6 Assessment Literacy

Although, at first, assessment appears to have been largely overlooked in CMC and language acquisition literature, it seems to be moving to the forefront of the debate on how to best evaluate telecollaborative practices. Lamy and Hampel (2007) cast this oversight in terms of development. The focus appears to have been, until recently, mostly on task design, media type, and philosophical frameworks: understandable, considering the relative novelty of the technology being incorporated.

However, Levy and Stockwell (2006) point out educators have failed to incorporate assessment methodologies that reflect the changing nature of course design. They claim that many educators are still using exams and tests at the completion of courses focused on telecollaborative designs. If blended learning is to be an integral part of language teaching, then it follows that assessment practices need to accurately reflect this pedagogical shift. Conceivably, there may be negative ramifications for course designers in which the learners themselves would start to question the legitimacy of assessment. O'Dowd (2010) highlights this issue in a timely article, where he states;

If educators believe that foreign language education in our modern 'globalized' society should involve the ability to learn, work and communicate in online contexts with members of other cultures, then it is to be expected that assessment procedures and criteria should take this new learning context into account (p. 338).

As O'Dowd goes on to point out, there are a range of complex issues connected to assessing the skills and competencies of language learners in a CMC-related course. Issues such as; Intercultural Communicative Competence (ICC), multimodalities and multiliteracies (Lamy and Hampel 2007), as well as interpretive skills and dialogue sensitivity (Schneider and von der Emde 2006).

Assessing these issues is problematic. Although O'Dowd (2010) makes some effort to describe current attitudes to, and methods for, assessing CMC courses, he describes aspects of (yet fails to focus on) what may be the main point of assessment: involving learners in the assessment process. Incorporating learners in the development of assessment rubrics as well as course design is an interconnected process. By getting learners to describe criteria for assessment, it can be inferred that they are undergoing learning of the key constructs and components of the course. Negotiating what construes what requires interlocutors (viz: learners and their instructors) to come to a common understanding and agreement before the assessment can be undertaken. Furthermore, it would be logical to undergo this process prior to the development of the material to be assessed.

8.5.7 Stakeholders

The main driving force behind co-operation and particularly, collaboration, is for stakeholders to help each other achieve objectives and this interaction is reciprocal in nature. In education however, it is recognized that instructors and students, although sharing certain objectives, do not usually share collegial and equitable control in their relationship. This control over the relationship usually becomes more defined the earlier the stage of education: elementary school students are usually told what to do, while graduate students often help their professors attain mutually beneficial objectives.

8.5.8 Methods

Most research seems to be focused on one or a mix of three methodologies: ethnological, discourse analysis or conversation analysis.

We need to consider what constitutes research data for either developing intercultural competence or language learning: output or interaction (Dooly 2011a). Although ethnographies, for example, provide rich detail of actual learning experiences, analysis of the data can suffer from segmentation that has implications for the validity and reliability of the research. One response to this conundrum has been to sidestep the perspective that knowledge should be tested and measured at the completion of a program and instead, focus on trying to capture and trace the emergence and evolution of students' learning moments throughout a course. This ethnographical approach has been labeled Activity Relevant Episodes (ARE) and was published in 2001 by Barab et al. (2001). The essence of this approach is to avoid preconceived ideas of what constitutes learning. One problem remains the same however, as Dooly (2011b) concludes ARE still lacks cross study validity.

Consequently some researchers try to focus on data that can provide cross study reliability. Discourse Analysis (DA) provides an obvious opportunity. It enables researchers to focus on a variety of language in any form produced by any number of users ranging from specific language types to corpus analysis. The fact that most language production in CMC environments is captured means DA is an ideal tool to analyze communication in empirical terms (Herring 2004). Initial efforts to apply DA used asynchronous tools, looking primarily at word counts and numbers of postings, but then moved toward a more semantic-based analysis of content (Fitzpatrick and Donnelly 2010), a reflection of both the ever growing multimodality of CMCs, as well as a move from cognitive analyses to SCT. Van Leeuwen (2008) argues that for DA to be effectively applied, researchers will need to move from a linguistic analysis to a more socio-semantic one, and not be shy of incorporating additional cultural theories to augment and inform DA.

One component of DA is Conversation Analysis (CA) which focuses on turn-taking, adjacency, and repair between interlocutors. CA was originally developed as a tool to analyze social interaction rather than language acquisition (Egbert et al. 2004; Hauser 2005; He 2004) although this has been challenged of late for use in conjunction with sociocultural and activity theories, situated learning theory, and longitudinal studies (González-Lloret 2011). Consequently for language acquisition, there are only a few studies (Kitade 2000, 2005; Negretti 1999; Thorne 2000; González-Lloret 2007, 2008, 2009) that analyze learners' foreign language acquisition. González-Lloret (2011) believes that for CA to demonstrate learning, expanding the definition of learning may be necessary, so that SLA is not limited only to linguistic features but also includes the social context and sequential development of interactions. In particular, the use of CA for the study of multimodal synchronous CMC is still relatively new (Jenks 2009). As interactional software becomes more sophisticated and internet connections become faster and more powerful, the use of video in connection with audio and text is becoming more common (González-Lloret 2011).

The value of being able to use CA and DA lies within the fact that authentic language is measurable in a classroom, as compared to traditional sources which have tended to be outside the classroom (González-Lloret 2011). Furthermore, being able to describe perceived changes in SL proficiency ensures that these methods provide valued insights for educational institutions as well as researchers.

8.6 Summary

Education is a huge service industry in most countries around the world. In the United States, it is second behind the health sector at 10 % of GDP (Larson 2009). It performs a number of roles in society – which, perhaps, accounts for its lack of change over the last 100 years (Larson 2009). This paper has covered just one aspect of the education sector – second language acquisition (SLA) in which one 'hard' technological innovation was examined: computer-mediated communication (CMC). Using Zhouying's perspective, whereby the notion of technology incorporates innovative methods of practice, thought and management in addition to normative concepts of artifacts, we can see that successful implementation actually requires advances in a *variety* of skills.

The 'soft-tech' in this paper included reviews of multimodal literacies, appropriacy of usage, and accessibility. It also included a look at various philosophical frameworks and how they have transformed CMC implementation over recent years. Finally, we looked at education management (blending, collaboration, curricular development, and international communicative competence – ICC), assessment literacy, and research methodologies. In this respect, at least, relevant researchers appear to be coming to grips with methods on how to effectively incorporate 'hard tech'.

Without question, Internet telephony has provided the mechanism by which many educational practices are being transformed. But, as described with the earlier reference to Vygotsky, it is not so much the shovel that can transform modes of production, it is how the shovel is *actually used* that can induce a revolution in technological practice. Likewise, practices and processes that incorporate IT-based functions need to be conceptualized as technological practices themselves. This is where we, as researchers, need to remember that the term technology should incorporate techniques, management approaches, and processes when we consider the ramifications of innovative artifacts.

References

Barab SA, Hay KE, Yamagata-Lynch LC (2001) Constructing networks of activity: an in-situ research methodology. J Learn Sci 10(1/2):63–112

Byram MS (1997) Teaching and assessing intercultural communicative competence. Multilingual Matters, Clevedon

Cheon H (2008) Sociocultural theory and computer-mediated communication-based language learning. TESOL Appl Linguist 8(1):1–3

Davison I (2012) $48 computer starts new digital revolution. http://www.nzherald.co.nz/business/news/article.cfm?c_id=3&objectid=10789863. Accessed 5 Mar 2012

Deardorff DK (2006) The Identification and assessment of intercultural competence as a student outcome of internationalization at institutions of higher education in the United States. Unpublished doctoral dissertation, North Carolina State University, North Carolina

Deutschmann M, Panichi L, Molka-Danielsen J (2009) Designing oral participation in second life – a comparative study of two language proficiency courses. ReCALL 21(2):206–226

Donato R (2000) Sociocultural contributions to understanding the foreign and second language classroom. In: Lantolf J (ed) Sociocultural theory and second language learning. Oxford University Press, Oxford, pp 27–50

Dooly M (2011a) Effective integration of telecollaboration: a teacher education initiative (colloquium paper, with Randall Sadler). In the changing context of globalization: American association of applied linguistics annual conference, Chicago, 26–29 Mar 2011

Dooly M (2011b) Divergent perceptions of telecollaborative language learning tasks: task-as-workplan vs. task-as-process. Lang Learn Technol 15(2):69–91

Egbert M, Niebecker L, Rezzara S (2004) Inside first and second language speakers' trouble in understanding. In: Gardner R, Wagner J (eds) Second language conversation. Continuum, London/New York, pp 178–200

Ellis R (2010) Foreword. In: Thomas M, Reinders H (eds) Task-based language learning and teaching with technology. Continuum, London, pp xvi–xviii

Falowo RO (2007) Factors impeding implementation of web-based distance learning. AACE J 15 (3):315–338
Fantini AE (2006) Exploring and assessing intercultural competence. Federation EIL, Brattleboro. sit.edu/SITOccassionalpapers/sitops01.pdf
Fitzpatrick N, Donnelly R (2010) Do you see what i mean? Computer-mediated discourse analysis. In Donnelly R, Harvey J, O'Rourke KC (eds) Critical design and effective tools for e-learning. IGI Global, Hershey, PA
Frater G (2000) Observed in practice. English in the national literacy strategy: some reflections. Read Lit Lang 34(3):107–112
Gereluk D (2005) Communities in a changing educational environment. Br J Educ Stud 53 (1):4–18
Gewirtz S (2002) The managerial school: post-welfarism and social justice in education. Routledge, London
González-Lloret M (2007) What do language learners attend to when their environment changes? In: Periñan C (ed) Revisiting language learning resources. Cambridge Scholars, Newcastle, pp 223–242
González-Lloret M (2008) Computer-mediated learning of L2 pragmatics. In: Soler EA, Martinez-Flor A (eds) Investigating pragmatics in foreign language learning, teaching and testing. Multilingual Matters, Clevedon, pp 114–132
González-Lloret M (2009) CA for computer-mediated interaction in the Spanish L2 classroom. In: Kasper G, Nguyen H (eds) Conversation analytic studies of L1 and L2 interaction, learning, and education. NFLRC and University of Hawaii Press, Honolulu, pp 281–316
González-Lloret M (2011) Conversation analysis of computer-mediated communication. CALICO J 28(2):308–325
Guth S, Helm F (eds) (2010) Telecollaboration 2.0: language, literacies and intercultural learning in the 21st century. Peter Lang, Bern
Hauck M (2010) Telecollaboration: at the interface between multimodal and intercultural communicative competence. In: Guth S, Helm F (eds) Telecollaboration 2.0: language, literacies and intercultural learning in the 21st century. Peter Lang, Bern
Hauser E (2005) Coding 'corrective recasts': the maintenance of meaning and more fundamental problems. Appl Linguist 26:293–316
He AW (2004) CA for SLA: arguments from the Chinese language classroom. Mod Lang J 88:568–582
Herring SC (2004) Computer-mediated discourse analysis: an approach to researching online behaviour. In: Barab SA, Kling R, Gray JH (eds) Designing for virtual communities in the service of learning. Cambridge University Press, New York, pp 338–376
Hodkinson H, Hodkinson P (2005) Improving schoolteachers' workplace learning. Res Papers Educ 20(2):109–131
Huckin T, Coady J (1999) Incidental vocabulary acquisition in a second language: a review. Stud Second Lang Acquis 21:181–193
Hulstijn JH (2003) Incidental and intentional learning. In: Doughty CJ, Long MH (eds) The handbook of second language acquisition. Blackwell, London, pp 349–381
Jenks C (2009) When is it appropriate to talk? Managing overlapping talk in multi-participant voice based chat rooms. Comput Assist Lang Learn 22:19–30
Kabata K, Edasawa Y (2011) Tandem language learning through a cross-cultural Keypal project. Lang Learn Technol 15(1):104–121
Kitade K (2000) L2 learners' discourse and SLA theories in CMC: collaborative interaction in internet chat. Comput Assist Lang Learn 13:143–166
Kitade K (2005) Interactional features of asynchronous computer-mediated communication for language learning: from cognitive and sociocultural perspectives. Unpublished dissertation, University of Hawai'i at Manoa, Honolulu
Kress G (2003) Literacy in the new media age. Routledge, London

Lamy M, Hampel R (2007) Online communication in language learning and teaching. Palgrave Macmillan, New York

Lantolf J, Thorne SL (2006) Sociocultural theory and the genesis of second language development. Oxford University Press, Oxford

Larson RC (2009) Education: our most important service sector. Serv Sci 1(4):i–iii

Levy M, Stockwell G (2006) CALL dimensions. Options and issues in computer-assisted language learning. Lawrence Erlbaum, Mahwah

Loveless A, Devoogd G, Bohlin R (2001) Something old, something new: is pedagogy affected by ICT? In: Loveless A, Ellis V (eds) ICT, pedagogy and the curriculum. Routledge, London, pp 63–83

Müller-Hartmann A, Schocker-v. Ditfurth M (2010) Research on the use of technology in task-based language teaching. In: Thomas M, Reinders H (eds) Task-based language learning and teaching with technology. Continuum, London

Negretti R (1999) Web-based activities and SLA: a conversational analysis approach. Lang Learn Technol 3(1):75–87

O'Dowd R (2010) Issues in the assessment of online interaction and exchange. In: Guth S, Helm F (eds) Telecollaboration 2.0. Peter Lang, Switzerland, pp 337–360

Ohta A (2001) Second language acquisition processes in the classroom: learning Japanese. Lawrence Erlbaum, Mahwah

Parker R, Ingram A (2011) Considerations in choosing online collaboration systems: functions, uses, and effects. J Res Center Educ Technol 7(1):2–15

Pegrum M (2009) From blogs to bombs: the future of digital technologies in education. University of Western Australia Press, Perth

Schneider J, von der Emde S (2006) Conflicts in cyberspace: from communication breakdown to intercultural dialogue in online collaborations. In: Belz JA, Thorne SL (eds) Internet-mediated intercultural foreign language education. Heinle & Heinle, Boston

Smith B (2003) The use of communication strategies in computer-mediated communication. System 31:29–53

Smith B (2005) The relationship between negotiated interaction, learner uptake and lexical acquisition in task-based computer-mediated communication. TESOL Q 39:33–58

Thorne SL (2000) Beyond bounded activity systems: heterogeneous cultures in instructional uses of persistent conversation. Paper presented at the 33rd Hawaii international conference on system sciences, Hawaii

Thorne SL, Lantolf JP (2007) A linguistics of communicative activity. In: Pennycook A, Makoni S (eds) Disinventing and reconstituting language. Multilingual Matters, Clevedon, pp 170–195

Timmerman CE, Kruepke KA (2006) Computer-assisted instruction, media richness, and college student performance. Commun Educ 55:73–104

Van Leeuwen T (2008) Discourse and practice: new tools for critical discourse analysis. Oxford University Press, New York

Warschauer M (1996) Telecollaboration in foreign language learning. University of Hawai'i Second Language Teaching and Curriculum Center, Honolulu

Wertsch JV (2002) Computer mediation, PBL, and dialogicality, special issue of Distance Education 23(1): 105–108

Wilhelm A (2011) How technology has changed education. http://thenextweb.com/insider/2011/01/05/how-technology-has-changed-education/. Accessed 15 Feb 2012

Zhouying J (2011) Global technological change: from hard technology to soft technology (2nd edn). University of Chicago Press, Chicago

Chapter 9
Transforming the Personal Response System to a Cloud Voting Service

Yu-Hui Tao and C. Rosa Yeh

9.1 Introduction

The personal response system (PRS), otherwise known as classroom (Fies and Marshall 2006), people (Griff and Matter 2008), student (Bunce et al. 2006), and audience (Bunz 2005) response system and electronic voting system (Draper and Brown 2004), has been increasingly used in classroom teaching since a new generation of infrared PRS became available in 1999 and subsequently used widely after 2003 (Abrahamson 2006). Figure 9.1 presents a graphical sketch of the PRS classroom environment. The instructor uses the teacher's infrared-based remote control to click a question from the PRS server, and the overhead projector projects the image of the question on a white screen in front of a class of students. The students can respond to the multiple-choice question using their own infrared-based remote controls, and see the confirmation of their click action; in this case, the changing color of their remote control numbers as displayed around the edges of the screen from green to red. All the click actions are immediately stored; thus, the teacher can show the statistics on screen to the students. Appropriate subsequent actions can then be taken from the perspectives of teaching instructions, which can be part of the pedagogical strategies designed for adopting PRS in the classroom setting.

In the West, PRS is commonly implemented in major universities, such as Harvard (ATG 2010), Cornell (CIT 2012), and Berkeley (ETS 2009). Why do these universities adopt PRS? According to the benefits summarized by the

Y.-H. Tao (✉)
Department of Information Management, National University of Kaohsiung, Kaohsiung,
Taiwan, R.O.C
e-mail: ytao@nuk.edu.tw

C.R. Yeh
Graduate Institute of International Human Resource Development, National Taiwan Normal
University, 162, Sec. 1, Ho-Ping E. Rd, Taipei 106, Taiwan, R.O.C
e-mail: rosayeh@ntnu.edu.tw

S. Uesugi (ed.), *IT Enabled Services*,
DOI 10.1007/978-3-7091-1425-4_9, © Springer-Verlag Wien 2013

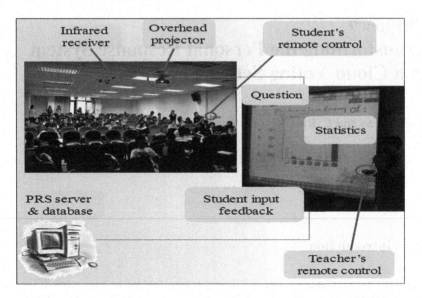

Fig. 9.1 PRS classroom facility

University of Wisconsin (Wiki 2012), PRS improves attentiveness, increases knowledge retention, polls anonymously, tracks individual responses, displays polling results immediately, creates an interactive and fun learning environment, confirms audience understanding of key points immediately, and gathers data for reporting and analysis. Taking Florida State University (FSU) as an example, Professor Calhoun demonstrated that the use of PRS increased student attendance from 60 % to more than 85 % and increased the size of the Economics class, resulting in the need for a larger classroom with 500 seats. Consequently, FSU extended the facility to ten lecture halls to cover other subjects, such as physics, economics, math, biology, geography, sociology, geology, nursing, and chemistry (Briggs 2006). Several universities in Taiwan have adopted the PRS technology, but Taipei Medical University (TMU) has the largest scale of implementation (http://www.wretch.cc/blog/habook/9560532) because it fulfilled the one-remote-control-per-student policy that approximates the ideal large-implementation approach of universities in the West (http://excellence.tmu.edu.tw/~TMU_TEACH/pro/super_pages.php?ID=pro1&Sn=20).

Although PRS technology has its pedagogical benefits in class interactions and instructions, its applications are still limited by physical location due to the requirement of specialized equipment, such as the ten lecture halls in the case of FSU and the student remote control policy in TMU. Internet technologies have started to mature, and have achieved a stage where many innovative, insightful, and interesting applications have become more feasible. Each technology has specific strengths and applications, and maximizes its potential may create new emergent applications with other technologies. PRS usage in college education is a good emergent candidate for such an evolving state for a wider scope of innovative applications.

Davila et al. (2006) classified innovations as incremental, semi-radical, and radical. One possible incremental innovation of PRS is freedom from its current applications using specialized equipment and fixed physical locations. PRS is an ideal candidate for transformation into a more widely used voting service. Instead of specialized clickers, personal mobile devices, such as cell phones, personal digital assistants (PDAs), pads, or notebook computers, of the audience/voters are used. This transformation potentially opens up the service to all synchronous or asynchronous group activities with voting needs.

With the integration of existing technologies, new IT-enabled services may be quickly created to meet the criteria of this incremental innovation. The remaining sections include a brief literature review on PRS, followed by the conceptual description of the proposed cloud voting system. Two research prototype applications are then introduced as proof of concept. Finally, a brief discussion of the new business service model by this improved PRS technology is discussed with a concise conclusion.

9.2 Literature Review

Before the application of PRS in education, this technology was first used in the military for filmed instruction materials in the 1960s, as investigated by Judson and Sawada (2002), who commented that even by today's technology standards, the early system was fairly sophisticated for educational use. For example, the Litton Student Response System reported in Boardman (1968) allowed five answers from A to E and provided feedback for correct response through vibrating buttons.

The use of PRS in class instruction has evolved from in-class reading quizzes to pre-class response and reading (Crouch and Mazura 2001) to increase the problem-solving skills of students (Levesque 2011) through peer discussion (Smith et al. 2009) or peer instruction (Crouch and Mazura 2001). As an ideal complement to peer instruction, just-in-time teaching is also used to help structure students' reading before class and to provide feedback for the instructor to tailor peer discussion questions to target student difficulties (Mazur and Watkins 2010). Meanwhile, the main target of large-class application has also been extended to small-class application (Smith et al. 2011) to enable the students to do the required reading before class, share their thoughts, and learn from their peers, as well as for the instructors to engage all students in the class.

The components of PRS have also evolved since then. First, PRS has changed from a specialized system to a web-based system (Carlson 2001), which is more accessible to the students and more manageable to the teachers. Meanwhile, the remote control has been improved from infrared-based to a better and more stable Radio Frequency Identification (RFID)-based alternative, which has been adopted by many suppliers, such as Turning Technologies (http://www.turningtechnologies. com/). Furthermore, specialized student remote controls have been replaced by the students' own cell phones (Carlson 2001), an innovation now available through PRS providers, such as Turning Technologies (http://www.turningtechnologies. com/studentresponsesystems/mobiledistancelearning/).

Recent studies provide adequate evidence on the positive perception of students on the use of PRS in higher education. Judson and Sawada (2002) provided the earliest summary of PRS use, where they investigated PRS use in the 1960s and 1970s, and then made recommendations. Modern PRS use after the 1970s to 2000 was also discussed, and corresponding findings were drawn. By contrast, Fies and Marshall (2006) reviewed the methods employed to assess PRSs, including pedagogical constructs in consideration of the traditional and next-generation PRS systems with 7 references from 1997 to 2004 and actual implementation studies with 14 references from 1996 to 2005. These two review papers provide valuable information for understanding the transition of PRS adoption from the early stage to this modern time.

Simpson and Oliver (2007) provided a summary of the pedagogical and organizational implications of PRS adoption, with corresponding perceptions of staff and students. In particular, they compared the practices of PRS use before 2000 and after 2000 and up to 2006, which bridged the knowledge of two eras of PRS development and practices. Meanwhile, Caldwell (2007) reviewed 25 peer-reviewed articles, which identified primary PRS users, articulated the rationale for PRS use, explored and questioned the strategies used by PRS, and identified its best practices. Caldwell's (2007) research is perhaps the most comprehensive review, with majority of its references published after 2000.

In referencing the abovementioned review articles, Kay and LeSage (2009) summarized PRS literature and identified 13 benefits and 12 challenges. The benefits of PRS are grouped into three categories, namely, classroom environment, learning, and assessment benefits. Classroom environment benefits include attendance, attention, anonymity, participation, and engagement. Learning benefits include interaction, discussion, contingent teaching, learning performance, and quality of learning. Assessment benefits include feedback and formative and comparative assessments. The challenges of PRS are grouped into three categories, namely, technology-based, teacher-based, and student-based challenges. Technology-based challenges include non-functional remote control devices and PRS. Student feedback, coverage, and question formulation are examples of teacher-based challenges. Student-based challenges include acceptability of new methods, discussion, effort, summative assessment, attendance for grades, identifying students, and negative feedback.

Kay and LeSage (2009) also posited key problems and future research directions from past PRS literature. Key problems encountered by current PRS research include the lack of systematic research methodology, bias toward using the anecdotal, lack of qualitative data, excessive focus on attitudes as opposed to learning and cognitive processes, and inconclusive samples derived from limited education settings. Four future research directions for PRS research were identified as follows: the need to determine why specific benefits and challenges influence PRS use, the need for an in-depth research that analyzes the impact of specific types of questions on creating a student-centered learning experience, the need for knowledge-rich learning that builds a classroom community that can facilitate the expansion of PRS to include social sciences subject areas and K-12 classrooms, and the necessity for more research on the individual differences in the use of PRS focusing on gender, year level, age, and learning style.

Several issues presented by these PRS literature provide evidence for this study. First, most PRS facilities require students to have clickers of their own, which may be purchased by the school and distributed to the class, or purchased/rented by the students themselves. These options are neither convenient nor cheap (Briggs 2006) for the schools and the students. Second, the clickers and receivers are usually infrared-based, which many studies have found to be unreliable, necessitating an upgrade to RFID-based devices (Murphy 2008). Owing to the limitations of infrared- or RFID-based clickers, the questions are limited to true-or-false or multiple-choice questions (Beuckman et al. 2006). Third, operating and debugging the PRS are extra burdens for the teachers (Hatch et al. 2005). Fourth, many PRS servers and receivers are installed in fixed locations in Taiwan. Thus, the technology is not applicable to general classrooms or other locations, as commonly experienced in Taiwan.

9.3   Concept of Cloud Voting System

PRS would be more useful if it can be used anywhere and anytime without being constrained by fixed locations. This leads to the concept of web-based, online, or Internet-enabled cloud services. Slightly different from Fig. 9.1, the cloud voting service depicted in Fig. 9.2 illustrates that the PRS server and database facilities can be accessed through the Internet via the teacher's notebook computer and students' remote controls, as well as cell phones, PDAs, pads, and notebook computers.

The implementation of a cloud voting service depends on the availability of student remote controls, which in general are purchased by the schools and distributed to and collected from students. To remove this tedious procedure in large-scale university implementation, students are required to rent, such as in Florida State University (Briggs 2006) or purchase, such as in University of West Florida (http://uwf.edu/its/instructionandresearch/classroomresponse.cfm), from the campus bookstores. In Taiwan, the only known large-scale implementing school is TMU, which adopted a different strategy by purchasing the student remote controls and loaning them for free to the students during their stay at Taipei Medical University (http://excellence.tmu.edu.tw/~TMU_TEACH/pro/super_pages.php?ID= pro1&Sn=20). Either way, PRS adoption depends on specialized student remote controls, which is always a barrier for teachers and schools. In an effort to solve this problem, Professor Junki of Erskine College in South Carolina negotiated a deal with local SPRINT PCS provider to loan them 200 free cell phones and a temporary cell phone tower on the campus (Carlson 2001). This problem also created opportunities for several service providers. According to "Audience Response"

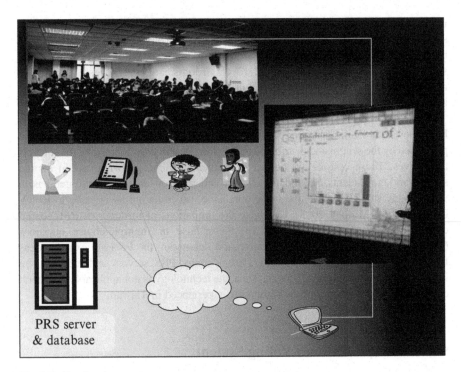

Fig. 9.2 Cloud voting service

Wiki (2012), there are already web-based PRS providers that use cell phones, such as eduVote (http://eduvote.de/), Your Emotion Live (http://www.yel.me), and Mentimeter (http://mentimeter.com), to provide a simplified free PRS service for everyone.

The feasibility of this cloud service also depends on whether the students have their own mobile devices. This is evident from the m-commerce applications which are "especially popular in Europe, Japan, South Korea, and other countries with strong wireless broadband infrastructure" (Laudon and Laudon 2011, p. 352). In Taiwan, a developing country, the majority of college students have cell phones and some have notebook computers. According to a 2010 investigation, nearly 80 % of junior high school students have cell phones and over 20 % of high school students have two cell phones (Huaxia 2010). Therefore, even in a developing country like Taiwan, it is promising for college students to access the cloud voting service via their own cell phones. Although only one-third of students may have 3G access through their cell phones, the Wi-Fi access in college campus is almost 100 %. Even outside the classrooms, the majority of college campuses in Taiwan have wireless connections built into every building. These connections can facilitate the voting services using Wi-Fi cell phones/notebooks or 3G cell phones. This new, unprecedented environment widens the application scope of voting services in campus activities.

9.4 Experimental Prototype

This concept of cloud voting system was implemented by a team of undergraduate students for their graduating project. During the 1-year duration of their project, five students proposed and implemented a simple version of cloud voting system for the 2010 Taipei International Flora Exposition (TIFE). This project was part of the requirements of the creative competition held by Chunghwa Telecom, along with a full version for using in college classrooms. More details are provided for the TIFE case because the mobile device is necessary in this setting. The full version, although more complicated, can be easily incorporated into university-level e-learning platforms, such as Moodle or Blackboard.

9.4.1 Cloud Voting System for 2010 Taipei International Flora Exposition

The objective of the TIFE project was to enable the PRS system to be available to tourists via wireless access to make voting-related activities a more instantaneous and fun experience. The core process of a cloud voting system for the TIFE activities, as illustrated in Fig. 9.3, is summarized as follows:

First, the TIFE-authorized person can log in to the system using a management account to set up questions and generate URL link for convenient access by tourists. Second, the voting URL is posted to the public so the tourists can access it from their own devices and register the ticket series number with their name and phone number for future identification and contact. Third, in addition to tourist access to the cloud voting system via the traditional desktop computers or notebook computers, the voting web pages have functional-oriented designs tailored for mobile devices for the user to easily submit the selected item to the server. Fourth, the TIFE office can check the voting results and statistical charts on the web pages in private or in public. The tourists who have voted can also check the outcomes after the voting activity ends. Fifth, the TIFE office can randomly draw the prize winners from those who voted or who voted specific answers.

Several activities can be created to take advantage of this cloud voting system. These activities could include calls for a prized quiz, quick question and answer, and the popular flower voting. This cloud voting system will not only save labor cost, such as tallying the vote statistics, and resource consumption, such as printing on papers, but can also increase interest and profit through this fresh and innovative way of interacting with tourists. To make voting convenient and eliminate concerns for compatibility issue for majority of the tourists, basic HTML approach can be used, which allows web pages to be downloaded through personal computers, notebooks, cell phones, PDAs, and other web-enabled mobile devices. To illustrate how the combination of the cloud voting system with TIFE tourists' own mobile devices can increase the variety of activities and attention of participants to create

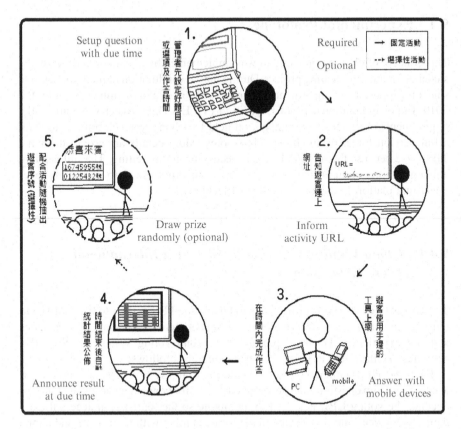

Fig. 9.3 Core process of the cloud voting system

word-of-mouth marketing, which is key to the successful application of the cloud voting system, some sample activities are presented in Tables 9.1, 9.2, 9.3, and 9.4. The activities are categorized according to time span, namely, extremely short, short, medium, and long.

These activities are mere references to the role of the cloud voting system in the TIFE, such that the office can design versatile activities to promote the TIFE by covering all possible types of tourists. For example, the biggest prize can be awarded to the tourists with tickets, and allow voting to be conducted at home or on-site with any device, on the road, or in any other place decided by the TIFE office for maximum accessibility. At the minimal, the application can be given to sponsors, such as Chunghwa Telecom, which provides prizes to customize voting activities for its own customers with Chunghwa Telecom mobile numbers. This method will form a bigger pool of prized-voting activities that could be participated by the tourists, the TIFE office, and the sponsors.

The TIFE voting system is named IRS@cloud, which can be a cloud-based or a web-based system, depending on the setup. IRS (Instant Response System) is another name for PRS. As seen in Fig. 9.4, users include the TIFE staff, activity

Table 9.1 Activities with extremely short duration (less than 15 min)

Activity	Description	Tactics	Frequency
Guided tour Q&A	During the guided tour, the guide can raise a quick question, such as knowledge on flowers. The tourists can immediately answer the question using their mobile device as an interactive game to bring up the fun atmosphere, as well as to test the level of understanding of the group	If the guide only faces a few tourists, the question can be announced verbally instead of setting up in the system beforehand	Depending on the volume of tourists 200 times/day
Quick response question	For large-scale prized-quiz activities, the TIFE office could gather the tourists in a large area with a stage or electronic board. The question is announced onsite for the audience to compete for the correct answer, which will heat up the cordial atmosphere	The audience competes for the correct answer based on their speed and correctness	Two to three times a week (medium prize value)
		Strong promotion beforehand to assure adequate number of onsite participants	
		Occasional higher prize will reinforce the retention of tourists	Two times a month (high prize value)

Table 9.2 Activities with short duration (15 min to 1 h)

Activity	Description	Tactics	Frequency
Daily flower king	All tourists with tickets can participate in this daily cosplay activity. Participants compete for the dressing-up votes of the audience to increase interest and fun		Once a day (top three winners)
Daily prized quiz	All tourists with tickets can participate in this daily prized quiz activity	The quiz questions are mainly based on the content of the TIFE exposition	Twenty times a day (Low-prize awards and one high prize at random times)
Up to you!	All tourists with tickets can participate in this daily game activity, wherein the winning tourist can request a task from a celebrity	Games can be designed to randomly draw a participant to play with a celebrity	Three times a day (the celebrity fulfills the task requested by the winners)

providers, and tourists. An activity provider is authorized to manage the voting activities, which includes setting the questions and response time and having the maintenance capability to change or delete activities. The simplest user are the

Table 9.3 Activities with medium duration (half day to 1 week)

Activity	Description	Tactics	Frequency
Flora star voting	Voting for the periodical TIFE Star. All winners will be competing for the TIFE Star in the last week of the exposition. Candidates are regular tourists and can be divided into male and female groups	One lucky tourist will be randomly drawn to have dinner with the elected TIFE Star	Once per half day to a week
Weekly prized quiz	Drawing the weekly prized quiz winner from all those who answered correctly in the weekly quiz. This voting activity can be done at home or in the exposition with mobile devices or voting facility provided by the TIFE office	Correct answers will be announced at the day of the activity Tourists with the correct answers may win the seasonal tickets	Once a week
TIFE thought sharing voting	Voting for the most insightful articles to encourage participants, particularly the students, to share their thoughts after visiting the TIFE park. Prices range from the top three best articles to ten distinguished articles	Best paper award Winning article will be posted in the official website and published in related publications Certificate and monetary reward	Once a week

Table 9.4 Activities with long duration

Activity	Description	Tactics	Frequency
Best tour guide voting	Voting for the best tour guide to encourage quality service and interactions with tourists	Top three winners Other voters may win smaller prizes	Once during the whole exposition
Monthly flower voting	Voting for the monthly most popular flower	Randomly drawn lucky voters who voted for the winning flower	Once per month
Champion flower voting	Voting for the tourists' favorite flower		Once during the whole exposition

tourists who can register and log in to the system, make queries on the voting activity, vote, make queries on the voting results, and leave messages. The most complex functionality is on the TIFE staff user who manages the membership and general data. The membership management role can add, change, and delete the accounts for tourists and activity providers, whereas the general data management

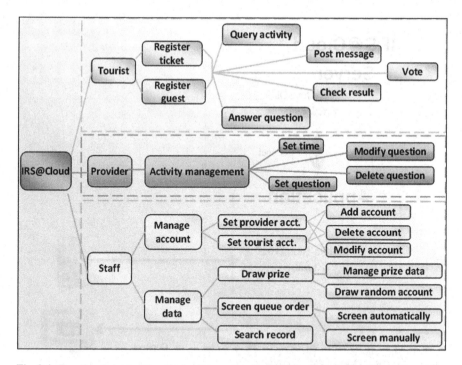

Fig. 9.4 Functional chart of the cloud voting system for the Taipei International Flora Exposition

role includes searching system records, screening the rank of incoming tourist time stamps, and drawing prizes.

The system development environment is depicted in Fig. 9.5. The server used is MySQL database under WinServer 2003. The application software used are PHP Web program language, Flash, XHTML, and JQuery. The application software used for mobile devices is limited to PHP and XHTML.

The following screen captures illustrate how the cell phone can complete a voting activity in the TIFE. The English translation is labeled next to the Chinese version to assist the reader in understanding these illustrations. The assumption is that any tourist with a ticket can register in the voting system because the ticket serial number is unique, as seen in Fig. 9.6. However, in reality, the TIFE did not design a serial number on the ticket, which has been officially criticized by several municipal councilmen. However, this condition does not affect the design of the application of this IRS@cloud system because there would have been a ticket serial number if this system was adopted by the TIFE Office in the beginning.

To confirm the identity of a prize winner, all voting participants are required to log in to the system, as seen in Fig. 9.7a. The participant should enter the activity serial number bound to the activity in any format accessible to the tourists, as seen in Fig. 9.7b.

In principle, all voting activities are multiple choice questions so the tourists can easily and quickly click one option and submit the answer back to the IRS@cloud

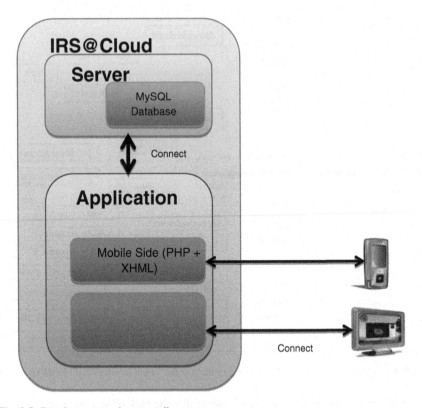

Fig. 9.5 Development environment diagram

Fig. 9.6 Cell phone/portable device interface: registering with ticket serial number

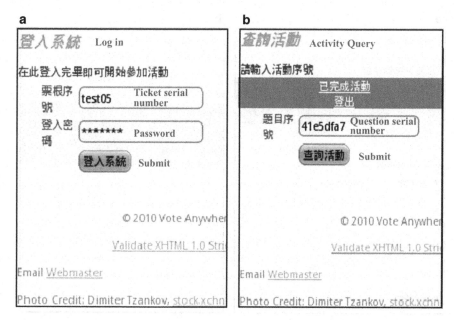

Fig. 9.7 (**a**) Logging in with the registered ticket serial number. (**b**) Entering the serial number of a voting activity

system, as seen in Fig. 9.8a. This procedure will be confirmed on the screen by a thank you message to the participating tourist, as seen in Fig. 9.8b.

9.4.2   Cloud Voting System for Classrooms

The objective of the classroom project is to enable the PRS system to become available to the university faculty and students via wireless access to conduct voting-related activities without being constrained by physical limitations.

There are three sets of facilities in the university attended by one of the authors. However, only a few teachers are using it, although many have expressed their interest in using this educational technology. One main reason for the teachers' hesitance in adopting PRS is their reluctance to change the location of the classrooms to be able to use the PRS functionality for a relatively small portion of the overall teaching activities, and manage the use of remote controls in class.

Complete installation of the PRS system software for a 50-seat classroom in Taiwan includes the PRS management software, one infrared receiver, one teacher's remote control, 50 student remote controls, and one remote control box. The entire system costs around US$4,430, excluding the server machine and operating software. Additional student remote controls are required for larger classes, which are expensive and hard to maintain. Furthermore, there are space

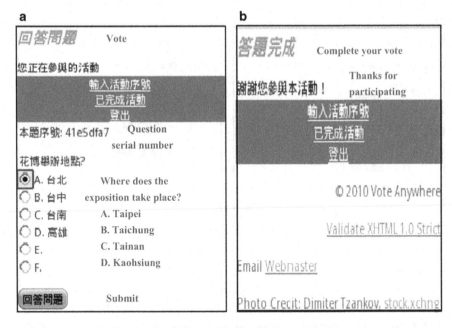

Fig. 9.8 (a) Selecting an answer option. (b) Completing the voting activity

requirements for the physical installation of PRS facility and the size of lecture room. Therefore, universities have to identify ideal classrooms for proper installation and future use. However, the management of remote controls, including distribution and collection in the classroom, is a big issue in practical implementation in Taiwan. Accordingly, although PRS functionality can be easily integrated into the Moodle e-learning platform, the student project requires a stand-alone PRS@cloud system to demonstrate its feasibility as a proof of concept.

Figure 9.9 presents the functional chart of PC and mobile devices used by students, as well as that of teachers and system administrator. Two modes are made available to students, namely, full-screen display using personal, notebook, or pad computers, or simplified screen display, such as mobile phones, personal digital assistants, or smaller pad computers. In either mode, students can register and login to the system, modify data, select classes, answer questions, and inspect the statistics. The teacher account can only be created by the system administrator, but once the teachers log into the system, they can create, inspect, and edit classes, as well as approve students, manage student name lists, add questions, and inspect statistics. The system administrator can create teacher accounts and maintain all accounts in the systems when necessary, such as when users forget their passwords.

This PRS function for educational use can actually be incorporated into the e-Learning platforms. The operation is no different from the general functions available in Moodle. In fact, the clicker provider, Turning Point Software, has released a Moodle module to support the clicker use with Moodle, but discontinued

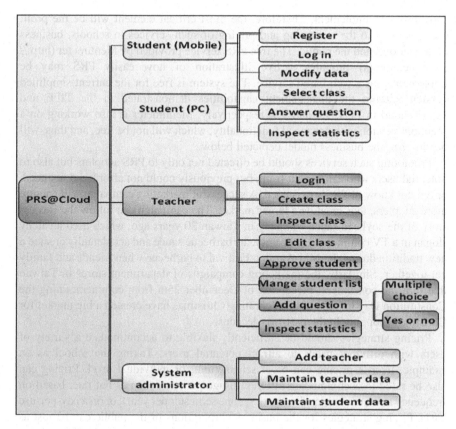

Fig. 9.9 PRS@Cloud class functional diagram

support for Moodle 2.x in January 2012 (http://moodle.org/mod/forum/discuss.php?d=45208). Therefore, this system is not demonstrated in this paper, particularly, the mobile phone operations are more complicated but similar to those screen captures as in the first case of TIFE above. However, with this PRS@cloud, whether stand-alone or integrated into the Moodle, there may be occasions when questions are asked for non-regular class activities, such as talks, events, and joint activities by more than two universities. A commercial PRS@Cloud service provider has to gradually expand and cover needs outside regular class activities.

9.4.3 Business Models and Implications

The general business model for cloud voting services will be any company that can provide such service for a fee. The infrastructure and the voting system can be self-developed or outsourced because it does not require challenging technologies and capabilities to implement. Therefore, the most critical element will be the profit model related to the marketing and pricing of such services to schools, business organizations, and the public. The free PRS service provided by Mentimeter (http://mentimeter.com) provides a live illustration of how easily PRS may be implemented as a web-based service. The system is free for the current simplified version without the management capabilities demonstrated in the TIFE and PRS@Cloud in Figs. 9.4 and 9.9, respectively. Mentimeter is also working on a premium version with enhanced functionality, which will not be free, and thus, will be closer to the business model depicted below.

Promoting such services should be directed not only to PRS adopters but also to potential users who can benefit from, but previously could not afford, did not need, or did not know of PRS's existence. A successful marketing campaign will capture more adopters, thus creating a larger market. The campaign can follow the success story of the soybean sauce company in Taiwan 20 years ago, which used a catchy slogan in a TV commercial to promote its barbecue sauce and accidentally created a new tradition during the Mid-Autumn Festival to barbecue when friends and family get together. Similarly, the marketing campaigns of department stores in Taiwan that turn the consumers' perception of December 25th from commemorating the Constitution Day in Taiwan to celebrating Christmas have created a big market for Christmas shopping in the past two decades.

Pricing strategies should be sufficiently flexible to accommodate a variety of users with different needs to attract potential users. Taking the school as an example, license pricing can be at school, unit, or individual level. Pricing can also be rated to accommodate different usage patterns, such as flat rate, based on frequency and time frame of usage (e.g., one semester per year), or on a pay-per-use basis. Pricing strategies for the business corporations or the public can be just as flexible.

Who can be the service providers? In Taiwan, there are currently only two local PRS suppliers who can easily adopt this new business model to transform their products and expand their service market. However, this business model can also be easily adopted by any corporation that wish to participate in cloud computing marketing, including brand-name notebook companies, such as Acer; telecommunication companies, such as Chunghwa Telecom, the largest in Taiwan; web portal companies, such as Yahoo and Google; and high-tech companies, such as Foxconn or Quanta.

For profitability concerns, the cloud voting service needs a larger market, not just for educational purposes. Traditional PRS application has been used in a broad range of industries and organizations, such as corporate training, game shows, delegate voting, and market research (Wiki 2012). Cloud voting services will expand its applications outside the sphere of the educational market. For example, the environment in large business organizations is more mature than university campuses. For small and medium enterprises (SME) without Wi-Fi capability, 3G cell phones remain the most feasible channel. These services may have been previously inaccessible, but the SMEs can begin using such voting services in public buildings or stores/shops, such as Starbucks or 7-11. Another potential application of cloud voting service is public events. Whether large or small, indoor or outdoor, audience voting in public events has become a trend. For example, New Year countdown shows, concerts, political gatherings, or government-held events usually last for hours and are attended by large crowds of audiences. Occasional audience interactions via cloud voting using their personal cell phones create an innovative and entertaining atmosphere within these events.

9.5 Conclusions

In this article, an incremental innovation of PRS has been described as an achievable, short-term technology integration for potentially profitable business opportunities. Although cloud voting service is a specialized service, it demonstrates how technologies can enable the transformation of a traditional product or service into a new business model. Furthermore, its influence can be sufficiently significant in changing the instructional design or classroom practices in different levels of the educational system. The influence of PRS may be sufficiently wide to be popularly used in many business routines that would have never considered before using such a service.

References

Abrahamson L (2006) A brief history of networked classrooms: effects, cases, pedagogy, and implications. In: Banks DA (ed) Audience response systems in higher education. Information Science, Hershey, pp 1–25

ATG (2010) Personal response systems (clickers). Harvard University. http://atg.fas.harvard.edu/icb/icb.do?keyword=atg&pageid=icb.page318552. Accessed 1 Mar 2012

Beuckman J, Rebello NS, Zollman D (2006) Impact of a classroom interaction system on student learning. In: McCullough L, Hsu L, Heron P (eds) Physics education research conference. American Institute of Physics, July 26–27, Syracuse, New York

Boardman DE (1968) The use of immediate response system in junior college. Unpublished master's thesis, University of California. Los Angeles

Briggs LL (2006) Response devices keep FSU students focused, Campus Technology. http://campustechnology.com/articles/2006/11/response-devices-keep-fsu-students-focused.aspx. Accessed 1 Mar 2012

Bunce DM, VandenPlas JR, Havanki KL (2006) Comparing the effectiveness on student achievement of a student response system versus online WebCT quizzes. J Chem Educ 83(3):488

Bunz U (2005) Using scantron versus an audience response system for survey research: does methodology matter when measuring computer-mediated communication competence? Comput Hum Behav 21:343–359

Caldwell JE (2007) Clickers in the large classroom: current research and best-practice tips. Life Sci Educ 6(1):9–20

Carlson S (2001) A professor uses cell phones in class to check students' comprehension. The Chronicle of Higher Education, June 21. http://www.erskine.edu/bq/chronicle.html. Accessed 1 Mar 2012

CIT (2012) Classroom polling, Cornell University. http://www.it.cornell.edu/services/polling/index.cfm. Accessed 1 Mar 2012

Crouch CH, Mazura E (2001) Peer instruction: ten years of experience and results. Am J Phys 69:970–977

Davila T, Epstein MJ, Shelton R (2006) Making innovation work: how to manage IT, measure IT, and profit from IT. Wharton School, Publishing, Upper Saddle River, NJ

Draper SW, Brown MI (2004) Increasing interactivity in lectures using an electronic voting system. J Comput Assist Learn 20(2):81–94

ETS (2009) Clickers (Audience response systems). U.C. Berkeley. http://ets.berkeley.edu/help/clickers-audience-response-systems. Accessed 1 Mar 2012

Fies C, Marshall J (2006) Classroom response systems: a review of the literature. J Sci Educ Technol 15(1):101–109

Griff ER, Matter SE (2008) Early identification of at-risk student using a personal response system. Br J Educ Technol 39(6):1124–1130

Hatch J, Jensen M, Moore R (2005) Manna from heaven or clickers from hell. J Coll Sci Teach 34(7):36–39

Huaxia (2010). Investigation of Taiwanese students with cell phones July 16. China Times News http://big5.huaxia.com/jjtw/dnsh/2010/07/1989168.html. Accessed 1 Mar 2012

Judson E, Sawada D (2002) Learning from past and present: electronic response systems in college lecture. J Comput Math Sci Teach 21(2):167–181

Kay RH, LeSage A (2009) Examining the benefits and challenges of using audience response systems: a review of the literature. Comput Educ 53:819–827

Laudon KC, Laudon JP (2011) Essentials of management information systems, vol 9. Pearson, Upper Saddle River

Levesque AA (2011) Using clickers to facilitate development of problem-solving skills. Life Sci Educ 10:406–417

Mazur E, Watkins J (2010) Just-in-time teaching and peer instruction. In: Simkins S, Maier M (eds) Just in time teaching: across the disciplines, and across the academy. Stylus, Sterling, pp 39–62

Murphy T (2008) Success and failure of audience response system in the classroom. In: Proceedings of the 36th annual ACM special interest group on University and college computing services fall conference. Portland, 19–22 Oct

Simpson V, Oliver M (2007) Electronic voting systems for lectures then and now: a comparison of research and practice. Aust J Educ Technol 23(2):187–208

Smith MK, Wood WB, Adams WK, Wieman C, Knight JK, Guild N, Sul TT (2009) Why peer discussion improves student performance on in-class concept questions. Science 323 (5910):122–124

Smith MK, Trujillo C, Su TT (2011) The benefits of using clickers in small-enrollment seminar-style biology courses. Life Sci Educ 10:14–17

Wiki (2012) Audience response. Wikepedia. http://en.wikipedia.org/wiki/Audience_response. Accessed 1 Mar 2012

Chapter 10
Case Studies of User Interface Design on Internet Banking Websites and Mobile Payment Applications in Thailand

Nagul Cooharojananone and Kanokwan Atchariyachanvanich

10.1 Introduction

In order to do payment transactions, customers conventionally have to go to a branch or a counter service of a bank, and so are restricted by location and service opening hours. However, this can be inconvenient for customers, for example due to traffic and limited servicing hours compared to their other commitments (such as work) in that limited service time window, or recent relocation to an unknown area. Therefore, many banks now increasingly use and offer web-based technologies to develop website access so that transactions can be done through the website anywhere and anytime. The concept of 24-h internet banking and equivalents has also been developed to cover mobile phones, since mobile phones are a large and continuously growing market and are served by an increasing number of mobile application developers, publishers and providers (Mobile applications 2008). Thus, they have become a business tool for companies to increase the opportunities to connect with customers. Accordingly, mobile payment (m-payment) applications have been created to assist in serving the customers since it is convenient for customers to use mobile phones to purchase goods and services and to transfer payments or to pay invoices for goods and services (http://www.wirelessintelligence.com/mobile-money).

In the case of university students, they are very busy with their studies and activities, but they still require to perform banking transactions, such as paying their registration fees, rent and bills. This is, however, analogous to other sectors of the community (for example with work commitments) and so is likely to be of a somewhat broader representation of the community. Therefore, for the general

N. Cooharojananone (✉)
Department of Mathematics and Computer Science, Chulalongkorn University,
Bangkok 10330, Thailand
e-mail: Nagul.C@chula.ac.th

K. Atchariyachanvanich
King Mongkut's Institute of Technology Ladkrabang, Bangkok 10520, Thailand

S. Uesugi (ed.), *IT Enabled Services*,
DOI 10.1007/978-3-7091-1425-4_10, © Springer-Verlag Wien 2013

populace, going to a bank in person to perform such transactions is not that convenient compared to internet banking (subject to local 24-h access to the internet). However, internet banking websites are different in design for each bank. For m-payment applications, although there are only a few applications currently available on the market, their designs are also different. These various different designs, for both the websites and mobile screens, could affect the ability of users (students in this study) to use them.

There have previously been several studies on user interface of banking service applications. Seven New Zealand online banks were studied in terms of their website's effectiveness, functionalities and internet strategies (Chung and Paynter 2002). Another study that focused on the trust, relative advantage and trialability of the application, found all three factors had a significant effect on the intention to use the application (Nor and Pearson 2007). In Thailand, there have been several reports on out of service-counter banking services. For example, the ATM interface of five banks was evaluated for their usability factors using four groups of users (Taohai et al. 2010; Cooharojananone et al. 2010), whilst for the internet banking services the quality of the services between each bank has been compared (Leelapongprasut et al. 2005; Ongkasuwan and Tantichattanon 2002; Esichaikul and Janeck 2009), but the range of factors that influence the customers intent to use the applications has not been extensively studied.

Previous research on mobile applications reported that the perceived ease of use has a positive effect on both the intention to use m-payment and the perceived usefulness of m-payment (Kim et al. 2010). Therefore, m-payment applications must be easy to understand and easy to use so as to increase the user's intention to use them. That is because the perceived usefulness will have a positive effect on the intention to use m-payment. For example, evaluation of the consumer's acceptance of mobile wallets found that trust aspect was related to the creditability of the m-payment and extended the user's acceptance to use the m-payment application (Kim et al. 2010). Likewise the degree of trust in virtual malls positively affected the customer's intention to use a mobile wallet. Moreover, good design aesthetics of the application can influence the intention of the user, where a higher level of design aesthetics of a mobile website will result in higher perceived levels of usefulness and ease of use of the mobile website (Shin 2009). Designers should consider having groups of potential users assist in the design by choosing the words and organizing the menus (Li and Yeh 2010) (Fig. 10.1).

The research reported here extends the previous research in internet banking and m-payments by evaluating eight and six usability factors, respectively, in compar-ing the interfaces and functions of two main systems for both internet banking and m-payments. Specifically, the goal of the research on the internet banking website and the m-payment application was to study the effect of the usability perspective on the intention to use. For internet banking, eight user factors (reliability, func-tionality, efficiency, ease of use, design aesthetic, learnability, satisfaction and security) were evaluated for their influence upon the intention to use the applica-tion. For the m-payment application, six factors (security, service quality, design

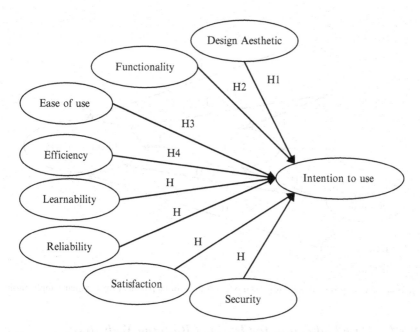

Fig. 10.1 The proposed hypotheses (H) model for the internet banking websites

aesthetic, trust and usefulness) were likewise evaluated. Both researches focused on university students as the participants.

10.2   Methodology

Each hypothesis (H) is based on the tenant that the given variable has a direct independent effect on the participant's intention to use the application. That is the design aesthetics (H1), functionality (H2), ease of use (H3), efficiency (H4), learnability (H5), reliability (H6), satisfaction (H7) and the security (H8) each have an effect on the participant's intention to use the application.

For the mobile payment applications, the hypotheses model is shown in Fig. 10.2, where in the same way as above each hypothesis (H) is based on the tenant that the given variable has a direct independent effect on the participants intention to use the application. That is the security (H1), service quality (H2) and design aesthetics (H3) each have an effect upon the user's trust of the application. In addition, the service quality (H4), trust (H5), design aesthetics (H6), usefulness (H7) and the ease of use (H8) each have an effect on the participant's intention to use the application.

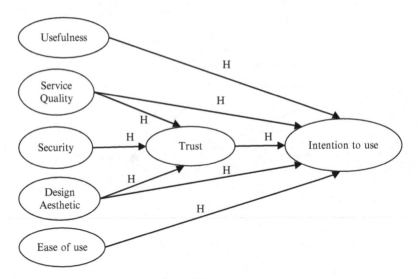

Fig. 10.2 The proposed hypotheses (H) model for the mobile payment (m-payment) applications

10.2.1 Data Collection for Internet Banking Websites

The research methodology on the internet banking websites was divided into two parts: (1) to survey the five most common tasks of internet banking, and (2) to use these to then test the websites with the participants (university students).

Our preliminary test was conducted to identify the top five banks and top five common tasks. To this end, 130 randomly selected university students, comprised of 38 males and 92 females, were asked to fill out a questionnaire. The questionnaire consisted of 40 questions that included personal information, their frequently used bank(s) and frequently used tasks. From the answers to these questions, the top five banks and the top five tasks were derived and are reported in Tables 10.1 and 10.2, respectively. The top five banks and top five tasks from Tables 10.1 and 10.2 were then to be used for the next experiment, to collect data for a regression analysis, but one of the top five tasks was excluded (see below). In this experiment 175 randomly selected university students (58 males and 117 females) were asked to fill out a new questionnaire. Their ages were between 18 and 25 years old. The questionnaire contained two sections to cover all eight perspectives. The first section was the demographic characteristics of the participants, whilst the second section was the participant's attitude towards each internet banking website. Within this second section were eight parts: to measure the reliability, functionality, efficiency, ease of use, design, learnability, satisfaction and security.

The participants were requested to evaluate the level of their agreement with each scale item on a four points Likert scale from disagree (1), somewhat disagree (2), somewhat agree (3), and agree (4). Questions in the questionnaire were selected from conventional works (Hornbaek 2006; Abran et al. 2003; Bevan 2001; Garvin 1987;

Table 10.1 List of the top five banks (for the sampled university students)

Rank	Bank
1	Siam commercial bank
2	Krungthai bank
3	Kasikorn bank
4	Bangkok bank
5	Bank of Ayudthaya

Table 10.2 List of the top five tasks (for the sampled university students)

Rank	Menu
1	Balance checking
2	Money transferring
3	Fee payment
4	Profile editing
5	Credit card payment

Kaikkonen et al. 2005). The tasks were then selected from the previously evaluated top five tasks (Table 10.2) except that for security reasons the credit card payment was not tested and therefore only the top four tasks were considered. The demographic data shows that all of the participants spend at least some time on the internet every day, with over one third spending 5–9 h per day on the internet (Table 10.3).

For each participant, they first selected one of the top five banks (from Table 10.1). Next, they had to select one of the top four tasks (Table 10.2 excluding the credit card payment) and try to complete it. Then, they completed the questionnaire. Examples of a captured screen of two internet banking websites are shown in Fig. 10.3.

10.2.2   Data Collection for Mobile Payment Applications

In this section, 200 randomly selected university students were split into two groups (100 participants in each). They were asked to perform one of the two tasks on the K-Mobile Banking PLUS application (http://www.kasikornbank.com) and one of the three tasks on the mPay application (http://www.mpay.co.th/). The common task between the K-Mobile Banking PLUS and the mPay applications was transferring money to another phone that uses the same payment service. However, the mPay application has two different tasks for transferring money; transferring money to a bank account or to mCASH (a money account from the service provider where that money is transferred from the user's bank account). The other task to be performed on the mPay application was a mobile top up, whilst the other task for the K-Mobile Banking PLUS system was payment of a bill. On completion of the tasks the participants were asked to fill out a new questionnaire appraising their views on the applications and their use.

The demographic data for both groups of 100 participants are shown in Table 10.4 with respect to the application the group they were in used. For both

Table 10.3 Demographic data of respondents (N = 175)

Category	Percentage (%)
Gender	
Male	33.1
Female	66.9
Frequency of using internet per day	
≤4 h	40.0
5–9 h	39.4
10–13 h	12.1
14–17 h	5.1
18–24 h	3.4

groups (i.e. for users of both applications) 61 % of the participants were female, most were 18–22 years old, and between 25 % and 34 % were flood victims from the 2011 flood in Thailand. The later is included as this is potentially a recent experience of when it may have been difficult to reach a bank service counter due to either flood water closure or access restrictions, or relocation to an unfamiliar location. Importantly, 87 % and 83 % of the subjects who used the K-Mobile Banking PLUS or the mPay application, respectively, had prior experience on using a touch screen mobile phone, whilst 20 % and 18 %, respectively, had prior experience with payment on a mobile phone.

For the K-Mobile Banking PLUS application (Fig. 10.4), the participant must enter a login name and password before they can start to do any tasks. The participant can then select the screen language from either Thai or English from the right of the main menu page.

For the mPay application (Fig. 10.5), the user interface of the main page is designed in graphics that consists of descriptions under each icon. But in the deep level of each task, such as the money transfer page (Fig. 10.5c), the user interface is not consistent but rather it has an appearance that is similar to a text-based design. The font size is very small forcing the normal (or corrected to normal) sighted users of this study to enlarge the screen to read the information and so would potentially be problematic for partially visually impaired users including uncorrected hyperopia (long sightedness) or those developing cataracts.

The questionnaire was comprised of two sections. The first section recorded the potential participant's demographic characteristics and their technology background, whilst the second section was the participant's attitude towards the M-pay application in terms of the (1) usefulness of application, (2) ease of use, (3) suitable aesthetic design, (4) trust in using the application, (5) service quality, (6) security and (7) intention of using the application. These questionnaires had 64 questions in total. All items were assessed using five-point Likert scales from strongly disagree (1), disagree (2), neutral (3), agree (4) and strongly agree (5). From our model (Fig. 10.2), we used the user interface, service quality, security and trust questions as the first study, and the usefulness, ease of use, user interface, service quality, trust and intention to use questions as the second study. Questions in the questionnaire were selected from conventional works (Kim et al. 2010; Shin 2009; Li and Yeh 2010; Klockar et al. 2003; Fred 1989).

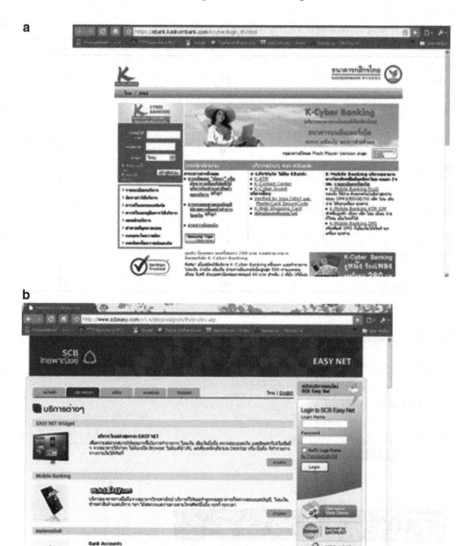

Fig. 10.3 Example of captured screen shots of the internet banking websites for (**a**) Kasikorn bank and (**b**) Siam commercial bank

10.3   Experimental Result

10.3.1   Internet Banking

Figure 10.6 shows the results of the multiple regression analysis of the results for the participant's intention to use the internet banking website. The beta-coefficients for the efficiency, ease of use, learnability and design aesthetics are shown in

Table 10.4 Demographic data of the respondents (N = 100 each)

Category	K-mobile banking PLUS (%)	mPay (%)
Gender		
Male	39	39
Female	61	61
Age		
18–22	84	96
21–35	7	4
≥35	9	0
Flood victims in Thailand		
Yes	29	34
No	71	66
Experience in a touch screen mobile phone		
Used	87	83
Never	13	17
Experience in mobile payment		
Used	20	18
Never used	80	82

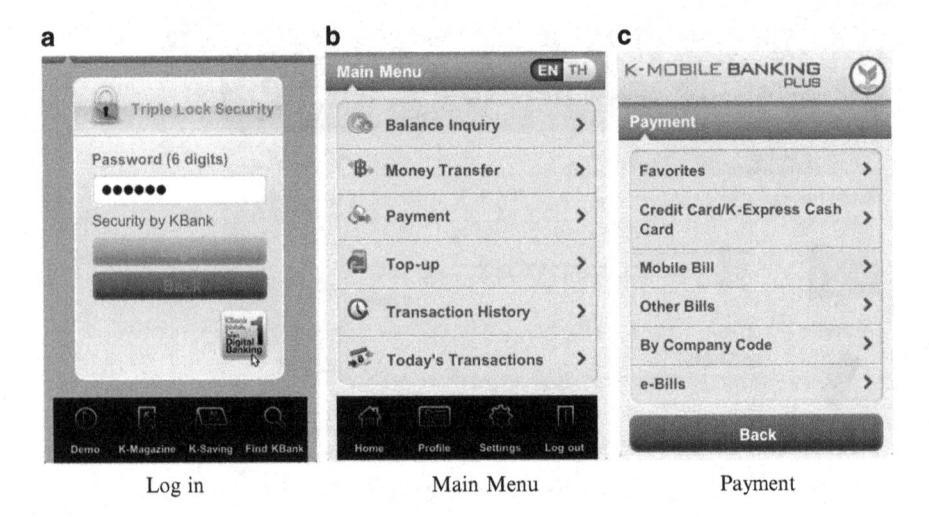

Fig. 10.4 Captured screens of K-mobile banking PLUS on iPhone

Fig. 10.6 and were all found to be significant (p < 0.01), and so hypotheses H1, H3, H4 and H5 are supported. In addition, the beta-coefficients for functionality and satisfaction were also significant, but at the p < 0.05 level, supporting hypotheses H2 and H7. In contrast, that for the reliability factor ($\beta = 0.006$; p > 0.05) and security ($\beta = 0.014$; p > 0.05) were not significant and so hypotheses H6 and H8 are rejected (and these are not shown in Fig. 10.6 accordingly).

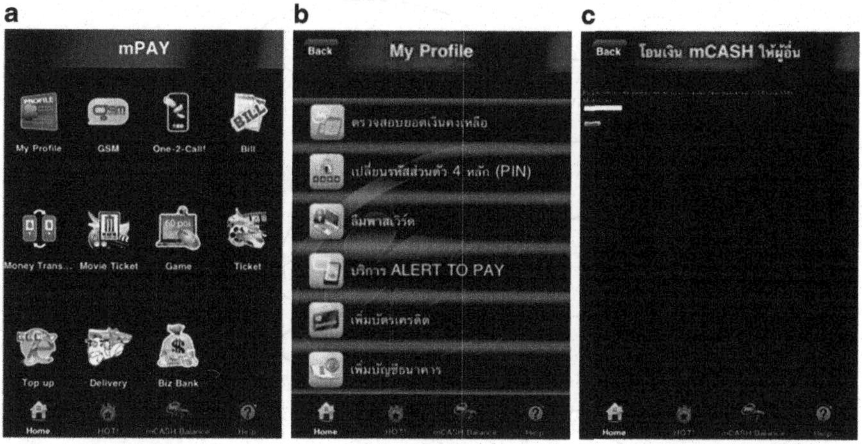

Fig. 10.5 Captured screens of mPay for money transferring task

That the reliability and security factors seem to have no significant effect on the participant's intention to use the application is because many of the participants think that these websites are official websites that have a good reputation and are safe and secure. Therefore, they intend to use these websites no matter how reliable they are as they believe that they are safe and secure. Another reason might be due to prior experience, as all the participants had no prior experience in internet banking and so may not have any ideas about security issues.

With respect to the six factors that did have a significant effect upon the participant's intention to use, the efficiency factor was mainly correlated with the intention to use factors, since the participants preferred to have a quick response and a high efficiency website. The ease of use factor had a high effect as participants commented that ease of use is an important key factor and is their main criteria in deciding if they intend to use any given internet banking website or not.

Functionality was important as participants were required to use a variety of different functions. Many internet banking websites provide common functions, such as balance checking and fee payment, but some provide functions that others do not provide, such as a calculator.

The learnability factor affected the intention to use factor as when participants had problems using the website, they looked for FAQ or the help menu for an example of how to do the tasks. In particular, fee payment was a difficult task for some participants. Thus, having help functions will enhance the user learnability and make the user continue to use the website.

The design aesthetic factor was also significant as participants agreed that a well designed website catches their attention and interest more. That might be one of the reasons why internet banking websites try to be different from each other in terms of their design. As a consequence, the satisfaction factor has an effect on intention to use factor.

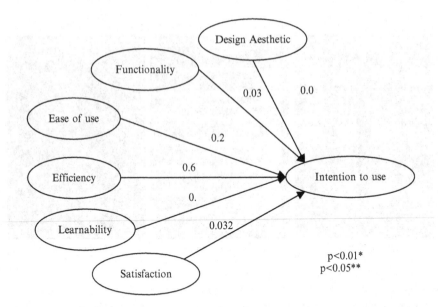

Fig. 10.6 The results for the internet banking website showing the beta coefficients for each significant factor (accepted hypothesis) that influences the intention to use the application

10.3.2 Mobile Payment

The experimental results for the mobile payment applications are considered separately as the results from (1) the regression analysis and (2) from the compared means analysis.

10.3.2.1 Result from Regression Analysis

Regression analysis was used to measure which of the three evaluated factors (user interface, service quality and security) had an effect on the feeling of trust in the application, and was also then used to measure which of the five evaluated factors (usefulness, ease of use, user interface, trust and security) had an effect on the intention to use the application. In this experiment, the collected data of the two tested systems, that of the K-Mobile Banking PLUS and the mPay applications, were analyzed separately. The two factors were paired and performed with multiple regression analysis, the results being shown in Figs. 10.7 and 10.8.

Figure 10.7 summarizes the results of the multiple regression analyses, showing the beta-coefficients and significance, for the intention to use the K-Mobile Banking PLUS application, including the R^2 and standardized path loadings for all hypothesized relationships. The feeling of trust in the application was significant, whilst the security, service quality and design aesthetics were all significant factors influencing the level of trust, supporting hypotheses H1, H2 and H3. Trust also

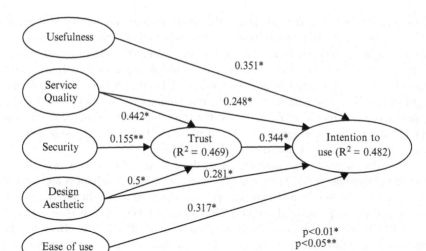

Fig. 10.7 The results of the hypotheses testing model for the K-mobile banking PLUS application, showing the beta-coefficient value and significance level for the accepted hypotheses

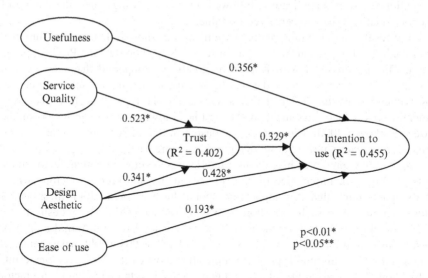

Fig. 10.8 The results of the hypotheses testing model for the mPay application, showing the beta-coefficient value and significance level for the accepted hypotheses

significantly impacted upon the intention to use this application, supporting hypothesis H5. Moreover, the service quality, design aesthetics, usefulness and ease of use were all found to significantly impact the intention to use the K-Mobile Banking PLUS application, supporting hypotheses H4, H6, H7 and H8, respectively. All together, the trust, service quality, design aesthetics, usefulness and ease of use accounted for 48.2 % of the variance in the intention to use the K-Mobile Banking PLUS application, with usefulness having the highest impact followed by (in decreasing order) the feeling of trust, ease of use, user interface and service quality. Figure 10.8 shows the results of the multiple regression analyses on the intent to use the mPay application, which revealed that the trust in the application is significant. The beta-coefficients for the service quality and user interface (design aesthetics) were significant factors upon the trust, supporting hypotheses H2 and H3, but security was not significant ($\beta = 0.102$; $p > 0.05$) and so hypothesis H1 is rejected (and accordingly not shown in Fig. 10.8).

Trust also significantly impacted upon the intention to use this application, supporting hypothesis H5. Moreover, the design aesthetics, usefulness and ease of use were found to significantly impact upon the intention to use the mPay application, supporting hypotheses H6 and H7. However, the service quality ($\beta = -0.060$; $p > 0.05$) was not significant with a slight numerical negative effect on the intention to use this mobile payment application, and so hypothesis H4 is rejected (and accordingly is not shown in Fig. 10.8). All together, the trust, service quality, design aesthetics, usefulness and ease of use accounted for 45.5 % of the variance in the intention to use the mPay application, with service quality having the highest impact on the level of trust and the user interface having the highest impact on the intention to use this application, followed by (in decreasing order) the feeling of service quality, trust and perceived usefulness.

The results from the multiple regression analysis showed that the user interface has the highest impact on the trust factor for the K-Mobile Banking PLUS application. This is consistent with Karvonen (2000), who reported that the aesthetic beauty of the website affects the feeling of trust in the internet environment. In Karvonen's study, the design elements, such as the clarity, was the most frequently mentioned key factor in enhancing the trust level towards service providers on the web (Karvoven 2000). Clear and clean design in the internet environment would make users ready to trust the service providers more easily.

In this study presented here the user interface is the most important factor and is then followed in importance by the service quality and security factors, where participants think that this application should have a confirmation message for finishing any tasks, such as a short messaging service (SMS). They desire that a SMS should be sent to their mobile phones after they have transferred money or paid bills confirming the transaction was completed as this would make them feel safer and trust in this mobile payment application more easily. Moreover, the results also show that the usefulness has the highest impact on the intention to use factor for the K-Mobile Banking PLUS application. Participants think that this application can be used in daily life for money transferring or any other tasks they want to do so that they do not have to go to the bank or the ATM.

The second highest impact on the intention to use factor is trust, which is an advantage of the K-Mobile Banking PLUS application. The application is owned by one of the top five banks in Thailand and so this familiar brand-name imposed level of trust relates to the level of user satisfaction. This is then followed by the ease of use and user interface, where because participants find this application easy to use

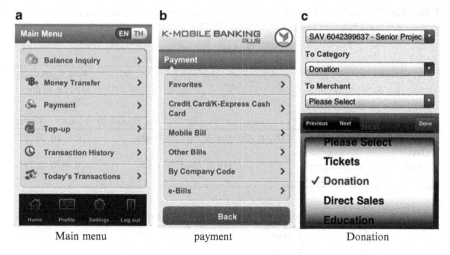

Fig. 10.9 Captured screen images of the bill payment task on the K-mobile banking PLUS application

they then concomitantly find the application is very convenient. Users also considered the user interface of the application, since a good design can allow them to be clearer in using the application and finish their tasks very quickly. Although the K-Mobile Banking PLUS application has a nice and consistent layout, most of participants who performed the bill payment task were confused about the menus because the location of its menu is different from the location on the ATM (Fig. 10.9).

The last factor was the service quality where some participants who failed to operate the application correctly then lost their trust of the application. In effect they perceived their errors as application errors and so the application becomes perceived by them as unreliable and untrustable.

10.3.2.2   Result from Compared Means

The results of compared means analysis are described for the difference between two groups. The point that we are interested is the potential significance of the participant's prior experience on using a touch screen mobile phone towards the ease of use and the intention to use. The results are shown in Tables 10.5, 10.6, 10.7 and 10.8.

(a) Is prior experience in using a touch screen mobile phone significant in the ease of its use?

Hypothesis:
μ_S = The average of ease of use for people who have experience of using a touch screen system on mobile phone

Table 10.5 The results of the compared means analysis for the K-mobile banking plus application

		Levene's test for equality of variances		t-test for equality of means					
								95 % confidence interval of the difference	
		F	Sig.	t	df	Sig. (2 -tailed)		Lower	Upper
Ease of use factor score	Equal variances assumed	0.009	0.926	0.463	96	0.645		−0.4710	0.7574
	Equal variances not assumed			0.478	14.552	0.640		−0.4970	0.7834

Table 10.6 The results of the compared means analysis for the mPay application

		Levene's test for equality of variances		t-test for equality of means					
								95 % confidence interval of the difference	
		F	Sig.	t	df	Sig. (2 -tailed)		Lower	Upper
Ease of use factor score	Equal variances assumed	0.239	0.626	−0.398	96	0.692		−0.6540	0.4357
	Equal variances not assumed			−0.417	22.280	0.681		−0.6516	0.4333

μ_M = The average of ease of use for people who never had experience of using a touch screen system on mobile phone

Null hypothesis (H0):
$\mu_S = \mu_M$: There is no significant experience difference in the ease of use.

Alternative hypothesis (HA): $\mu_S \neq \mu_M$: There is a significant experience difference in the ease of use.

Assumed that:

1. Dependent variable "ease of use" is a numerical variable.
2. The dependent variable is normally distributed.
3. The two groups have approximately equal variance on the dependent variable.

The independent samples t-test was used because it tests whether the means of two groups (experienced and non-experienced groups) are equal or not, and the means in this test are numerical variables (ease of use).

In Table 10.5, the Levene's test for equality of variances shows that the F ratio (0.009) is not significant (p = 0.926), and so the two variances are not significantly different. Thus, equal variances were assumed with a t value of 0.463 and 96

Table 10.7   The results of the compared means analysis for the K-mobile banking plus application

| | | Levene's test for equality of variances | | t-test for equality of means | | | | |
| | | | | | | | 95 % confidence interval of the difference | |
		F	Sig.	t	df	Sig. (2 -tailed)	Lower	Upper
Intention to use factor score	Equal variances assumed	0.327	0.569	0.480	98	0.632	−0.4491	0.7357
	Equal variances not assumed			0.473	15.644	0.643	−0.5005	0.7871

Table 10.8   The results of the compared means analysis for the mPay application

| | | Levene's test for equality of variances | | t-test for equality of means | | | | |
| | | | | | | Sig. | 95 % confidence interval of the difference | |
		F	Sig.	t	df	(2-tailed)	Lower	Upper
Intention to use factor score	Equal variances assumed	0.279	0.599	−0.301	98	0.764	−0.61135066	0.45013429
	Equal variances not assumed			−0.288	22.060	0.776	−0.66191484	0.50069847

degrees of freedom. The obtained p value (0.645) is not greater than the 95 % confidence intervals and the calculated t value (0.463) does not exceed the table t value of 1.960. Therefore, the null hypothesis could not be rejected and the two means (μ_S and μ_M) are not statistically significantly different at the 5 % level of significance. There is no significant experience difference in the ease of use for the K-Mobile Banking PLUS application.

In Table 10.6, the Levene's test for equality of variances shows that the F ratio (0.239) is not significant (p > 0.05) and the two variances are not significantly different. Thus, equal variances were presupposed with a t value of −0.398 and 96 degrees of freedom. The obtained p value (0.692) is greater than the 95 % confidence interval and the calculated t value (−0.398) does not exceed the table t value of 1.960. Therefore, the null hypothesis could not be rejected and the two means (μ_S and μ_M) are not statistically significantly different at the 5 % level of significance. There is no significant experience difference in the ease of use for the mPay application. That prior experience of using a touch screen system on a mobile phone

Table 10.9 The list of advantages and disadvantage of the mPay application

Advantages	Disadvantages
1. Application screen is graphic-based	1. The icons don't allow the user to immediately identify their function
2. Application form uses icon images with an underlying description	2. The graphic-based screen takes longer to load than a text-based one
3. Has full functionality in terms of paying for utilities and entertainment	3. Have to connect to the internet
4. Makes daily life easier and more convenient	4. Cannot customize icons for other users
5. Screen by the icon organizes the work into easy to use categories	5. Fee for service is more expensive than other applications
6. Frequently used functions can be added to the favorite menu, which makes management of applications consistent with the individual	6. Can only be used through the AIS, GSM and 1–2-call networks
7. Good response time and a clear status indication while the transaction is underway	7. Must have a mCASH account
	8. Registration is difficult and each bank has a different system

does not have any effect on the ease of use for mobile payment applications agrees with the participants who said that both the K-Mobile Banking PLUS and the mPay applications have transaction tasks that are useful and easy to use.

(b) Is prior experience on using a touch screen mobile phone significant in determining the intention to use?

Hypothesis: $\mu_S =$ The average intention to use for people who have prior experience on using a touch screen system on a mobile phone.

$\mu_M =$ The average intention to use for people who have no experience of using a touch screen system on a mobile phone.

Null hypothesis (H0): $\mu_S = \mu_M$: There is no significant experience difference in the intention to use.

Alternative hypothesis (HA): $\mu_S \neq \mu_M$: There is a significant experience difference in the intention to use.

Assumed that:

1. Dependent variable "intention to use" is a numerical variable.
2. The dependent variable is normally distributed.
3. The two groups have approximately equal variance on the dependent variable.

The independent samples t-test was used because it tests whether the means of two groups (experienced and non-experienced groups) are equal or not, and the means in this test are numerical variables (intention to use).

In Table 10.7, the Levene's test for equality of variances showed that the F ratio (0.327) is not significant (p = 0.569), and so the two variances are not significantly different. Thus, equal variance was assumed with a t value of 0.480 and 98 degrees of freedom. The obtained p value (0.632) is not greater than the 95 % confidence

Table 10.10 The advantages and disadvantages of the K-mobile banking PLUS application

Advantages	Disadvantages
1. Able to make financial transactions easier and simpler by the use of mobile phones	1. The menu button looks like it is text based which confuses the user
2. Saves time for financial transactions	2. Can change the language, but users must first access the settings in the Thai menu
3. The functions are essentially the same as in the K-ATM	3. The magazine page format is different from other pages (Looks like a webpage, not an application page.)
4. Screen is simple and uncomplicated	
5. Text font for easy reading (text base)	
6. Have video clips that help make it more user friendly	
7. Perceived security of password controlled access	
8. Records previous transactions	
9. Account shows previous transfers automatically in the transfer menu	
10. Able to use with every mobile telecom network	
11. Good classification of transactions	
12. You can find the nearest ATM machine from the GPS network	
13. The Kasikorn bank logo is in every page to make sure that the user does not accidentally log out	

interval, and the calculated t value (0.480) does not exceed the table t value of 1.960 and so the null hypothesis could not be rejected and the two means (μ_S and μ_M) are not statistically significantly different at the 5 % level of significance. There is no significant experience difference in the intention to use the K-Mobile Banking PLUS application. In Table 10.8, the Levene's test for equality of variances shows that the F ratio (0.279) is not significant (p = 0.599) and the two variances are not significantly different. Thus, equal variances were presupposed with a t value of 0.301 and 98 degrees of freedom. The obtained p value (0.764) is greater than the 95 % confidence intervals and the calculated t value (0.301) does not exceed the table t value of 1.960. Therefore, the null hypothesis could not be rejected and the two means (μ_S and μ_M) are not statistically significantly different from zero at the 5 % level of significance. There is no significant experience difference in the intention to use the mPay application.

Consistent with this analysis, that prior experience of using a touch screen mobile has no difference on the participant's intention to use the application, is that the participants mentioned that both applications were easy to use. The user interface of the two applications is consistent and not complex, so they are comfortable using these applications as they allow the participants to perform the tasks easily.

10.3.2.3 Results of the Participants' Comment

In this section, we summarize the advantages and disadvantages of the two systems, as derived from analysis of the participants' comments after completing the tasks. Table 10.9 shows the results for the mPay application whilst Table 10.10 shows that for the K-mobile banking PLUS application.

10.4 Conclusion

In this research, we studied the user interface design factor on internet banking websites and mobile payment applications in Thailand. For the study of internet banking websites, the application efficiency, satisfaction, functionality, ease of use, learnability and design aesthetics were all significant influences on the intention to use, whilst the reliability and security had no effect. For mobile banking applications, the service quality and design aesthetics affected the trust in the application, whilst the trust, service quality, usefulness, ease of use and design aesthetics all significantly affected the intention to use the application. In contrast, the security had no significant effect on the trust and the service quality had no effect on the intention to use. Due to the ease of use and learnability of the applications, prior experience of touch screen mobile phones had no effect on the ease of use and intention to use.

There are many more interesting factors that can be explored in future work, such as the literacy and loyalty factors. This research would be beneficial for any providers.

References

Abran A, Khelifi A, Suryn W, SeffahAbran A (2003) Usability meanings and interpretations in ISO standards. J Softw Qual 11:325–338

Bevan N (2001) International standards of HCI and usability. Int Hum Comput Stud 55:553–552

Chung W, Paynter J (2002) An evaluation of internet banking in New Zealand. In: Proceedings of the 35th Hawaii international conference on system sciences, Hawaii, pp 2410–2419

Cooharojananone N, Taohai K, Phimoltares S (2010) A new design of ATM interface for banking services in Thailand. In: Proceedings of the 10th annual international symposium on applications and the internet, Izmir, pp 312–315

Davis Fred D (1989) Perceived usefulness, perceived ease of use, and user acceptance of information technology. MIS Quart 13(3):318–323

Esichaikul V, Janeck P (2009) A survey of e-banking performance in Thailand. J Electron Financ 3:353–373

Garvin DA (1987) Competing on the eight dimensions of quality. Harvard Bus Rev 65(6):101–109

Hornbaek K (2006) Current practice in measuring usability: challenges to usability studies and research. Int J Hum Comput Stud 64:79–102

http://www.kasikornbank.com

http://www.mpay.co.th/

http://www.wirelessintelligence.com/mobile-money

Kaikkonen A, Kallio T, Kekalainen A, Kankainen A, Cankar M (2005) Usability testing of mobile application: a comparison between laboratory and field testing. J Usability Stud 1:4–16

Karvoven K (2000) The beauty of simplicity. In: Proceedings of the 2000 ACM conference on universal usability, Arlington, VA, pp 85–90

Kim C, Mirusmonov M, Lee I (2010) An empirical examination of factors influencing the intention to use mobile payment. Comput Hum Behav 26(3):310–322

Klockar T, Carr DA, Hedman A, Johansson T, Bengtsson F (2003) Usability of mobile phones. In: 19th international symposium on human factors in telecommunication, Berlin, pp 197–204

Leelapongprasut P, Praneetpolgrang P, Paopun N (2005) A quality study of internet banking in Thailand. In: Proceedings of the fourth international conference on ebusiness, Samutprakarn, Thailand, pp 61–65

Li Y-M, Yeh Y-S (2010) Increasing trust in mobile commerce through design aesthetics. Comput Hum Behav 26(4):673–684

Mobile applications (2008) The Mobile Marketing Association (MMA), pp 1–9

Nor KM, Pearson J (2007) The influence of trust on internet banking acceptance. J Internet Bank Comm 12(2):1–10

Ongkasuwan M, Tantichattanon W (2002) A comparative study of internet banking in Thailand. In: Proceedings of the first national conference on electronic business, Bangkok, Thailand, pp 24–25

Shin D-H (2009) Towards an understanding of the consumer acceptance of mobile wallet. Comput Hum Behav 25(6):1343–1354

Taohai K, Phimoltares S, Cooharojananone N (2010) Usability comparisons of seven main functions for Automated Machine (ATM) banking Service of five banks in Thailand, In: Proceedings of the international conference on computational science and its applications, Fukuoka, Japan, pp 176–182

Chapter 11
Japanese Students' Behavior Toward E-Commerce

Takashi Okamoto and Nobuyuki Soga

11.1 Introduction

With an increase in the use of information and communication technology (ICT) devices and the broadband environment in Japan, more diverse types of people are accessing various affordable, high-speed Internet services. The expansion of ICT affects our daily lives, giving us new options for purchasing goods and services. For example, we can easily obtain specific information on goods and services from the Internet and read other buyers' reviews. With such benefits, purchasing goods and services from online shops has become common.

Therefore, Internet selling has spread; even people with little knowledge of ICT find it easy to sell their goods at online malls and auction sites. In the ICT environment, we can consider electronic commerce (EC) as IT-enabled services and business-to-consumer (B2C) sales as a common EC service.

Japan's B2C market was worth 7.8 trillion yen in 2010, up 16.3 % from the previous year. The EC ratio, a measure of EC expansion, increased to 0.4 points from the previous year to reach 2.5 % (Ministry of Economy, Trade and Industry2010). As this report notes, commercial transaction computerization has been evolving and growing. Another report shows that 53.6 % of Internet users used B2C in 2010 (Ministry of Internal Affairs and Communications, Japan 2011a). In Japan, 46.9 % of Internet users indicated "purchasing goods and services" as their purpose for using the Internet via a PC, and 30.1 % indicated the same purpose for using their mobile phone (Ministry of Internal Affairs and Communications, Japan 2011b). This shows that B2C is commonly used as an important element of the consumer market, not only in Japan but worldwide.

Although B2C shops are not always profitable, their number has been increasing. In fact, only 32.2 % of online shops make profit, 46 % suffer losses, and 25.5 %

T. Okamoto (✉) • N. Soga
Ehime University, Matsuyama, Ehime Pref., Japan
e-mail: tokamoto@LL.ehime-u.ac.jp; soga@ehime-u.ac.jp

S. Uesugi (ed.), *IT Enabled Services*,
DOI 10.1007/978-3-7091-1425-4_11, © Springer-Verlag Wien 2013

177

expect the difficult business environment to continue (Nikkei Inc. 2010). Many online shops' business problems should not be attributed to technological complexities but to their management's lack of knowledge and knowhow. Many online shop managers do not understand online consumer behavior and their perceptions regarding online shops. Because small and micro enterprises and sometimes individuals manage their own online shops, they do not know how or have the resources to research the B2C market.

Several reports have profiled consumer behavior in online shopping. For example, the Japan Direct Marketing Association (JADMA) reports the state of consumer utilization of direct marketing on the Internet (The Japan Direct Marketing Association 2010). Its respondents, however, are limited to people who either have been using online shops or are familiar with the Internet, and thus, are more likely to patronize online shops. To develop online shops' market and create effective sales strategies, shop owners/managers need to survey users', including potential users', behaviors, purposes, and perceptions.

Regarding Internet use, the under-40 age group's Internet usage rate exceeds 96 % (Ministry of Internal Affairs and Communications, Japan 2011b). Because nearly all young people use the Internet, the online shop market will expand in the near future. There are few studies, however, on consumer behavior and perceptions of people aged 20 and below. Because they are major customers of online shops, understanding their behavior and perceptions is necessary for developing sales strategies for the online shop market.

In this study, we examine young people's use of ICT and the Internet and their behavior and perceptions regarding online shops. In particular, we report the characteristics of high school students and university students on the basis of our research. Conjoint analysis reveals their priorities with respect to several features of online shops. Our study will contribute to the development of the online shop market and suggest effective marketing strategies for online shops.

11.2 Summary of Research Model

11.2.1 Research Methods

To examine young people's behavior and perceptions regarding online shops, we use a questionnaire as the research method. We choose high school and university students, around 20 years old, as representative of young people. Our research was conducted on October 7, 2010 at Ehime University and November 18, 2010 at Ehime University senior high school. As both schools are located in the same area, we can reduce the number of demographic attributes other than high school or university age grouping.

11.2.2 Overview of Questionnaire

Our questionnaire is composed four sections. The first gathers information about students' state of using ICT devices and Internet services. The questions deal with the following topics.

- Possession of PC or mobile phone
- Duration of using Internet services on weekdays/weekends via PC/mobile phone (excluding e-mail)
- Internet services used

The second section gathers information about students' usage of online shops.

- Experience of using online shops: mainly from PC, mainly from mobile phone, equally from both PC and mobile phone, or no experience
- Categories of goods that students have previously purchased or expect to purchase from online shops (multiple answers accepted): "Book," "CD," "DVD," "Clothes," "Accessories," "Bag," "Shoes," "Watch," "Miscellaneous goods," "Purse/wallet," "Games," "Foods/drinks," "Cosmetics/perfume," "Sporting goods," "Music," "Instrument/score," and "Others"
- Priorities of the advantages of online shopping (multiple answers accepted): "Purchasing goods not sold at nearby physical shops," "Shopping at home," "Price/cheapness," "Large variety of items," "Open 24 h," "Referring reviews and reputations," "Purchasing at one's own pace," "Delivery by other people," "Easy searching of goods," "Easy comparison of goods," "Rich information about goods," "Easy purchasing procedure," "Rich information about other goods," "Availability of rare goods," "Speedy purchasing," "Acceptable waiting time for delivery," and "Others"
- Unattractiveness and anxieties regarding online shops (multiple answers accepted): "We cannot confirm goods' quality or details before purchasing," "Risk of difference between goods' online appearance and that of actual goods," "Risk of receiving inferior goods," "Annoying advertisements after purchasing," "Risk of personal information theft," "Risk of nondelivery of purchased goods," "Difficulty in canceling purchase orders," "Too long from purchase to delivery," "Complex purchasing procedure," "No supply of goods in stock at many online shops," "Too many items to recognize what I want," and "Others"

Because our research focused on online shops that sell goods, we eliminated online shops that sell services such as online reservations or music downloads. We referred to JADMA's research (The Japan Direct Marketing Association 2010) for selecting each option, and we transformed options so that they were appropriate for high school and university students.

Third section gathers data about students' perceptions of online shops, using conjoint analysis.

Last section gathers data about students' demographic attributes as follows:

- Transportation to their school
- Monthly budget
- Grade
- Gender

11.3 Features of Respondents

11.3.1 Lifestyle and Budget

The respondents were 350 high school students and 157 university students. Table 11.1 displays their composition, which contains only minor deviations in grade and gender, although the number of university students is half that of high school students.

Ninety percent of high school students and 70 % of university students ride their bicycle to school, and only 5 % of all students use public transportation to get to school. Because Matsuyama city, in which both schools are located, is compact, most students do not need public transportation. Very few students in Ehime use ICT devices while traveling in trains or buses, although students in urban area often use ICT devices while traveling in public transport. We can assume that they use ICT devices primarily at their house or school.

The monthly average budget of high school students is approximately 5,300 yen, and 67 % of high school students' budget is 2,000–6,000 yen. In contrast, the monthly average budget of university students is approximately 36,000 yen, and 45 % of university students' budget is 10,000–30,000 yen.

11.3.2 State of ICT Usage

Most of our respondents use both ICT and the Internet, as indicated by other research. Only 0.6 % of university students and 6.9 % of high school students do not have a PC at their homes, and only 0.6 % of university students and 3.1 % of high school students do not have a mobile phone. Thus, the vast majority of these young people have an ICT environment including both PCs and mobile phones.

Table 11.2 shows high school students' daily Internet access time (excluding e-mail). On weekdays, most of them access the Internet for 0.5–1 h or do not access it at all. They use the mobile phone as their primary device for accessing the Internet. Approximately 20 % of them use the Internet for 1–3 h on a weekday. For both PCs and mobile phones, the daily access time on weekends is longer than that on weekdays.

Table 11.1 Research methods and participants

Date	11/18/2010	10/7/2010
Method	Self-administered questionnaires (held and collected in class)	
Respondents	Ehime Univ. senior high school	Ehime Univ.
Number	350	157
Grade	1st:120	2nd:67
	2nd:116	3rd:66
	3rd:114	4th:24
Gender	Female:153	Female:77
	Male:197	Male:80

Table 11.2 High school students' daily Internet access time

	PC		Mobile	
	Weekday	Weekend	Weekday	Weekend
0 h	115 (33 %)	79 (23 %)	79 (23 %)	77 (22 %)
0–0.5 h	85 (24 %)	65 (19 %)	92 (26 %)	65 (19 %)
0.5–1 h	79 (23 %)	76 (22 %)	83 (24 %)	72 (21 %)
1–3 h	59 (17 %)	91 (26 %)	79 (23 %)	82 (23 %)
3–5 h	11 (3 %)	29 (8 %)	12 (3 %)	38 (11 %)
5 h over	0 (0 %)	9 (3 %)	5 (1 %)	16 (5 %)
No answer	1 (0 %)	1 (0 %)	0 (0 %)	0 (0 %)

Table 11.3 University students' daily Internet access time

	PC		Mobile	
	Weekday	Weekend	Weekday	Weekend
0 h	10 (6 %)	7 (4 %)	13 (8 %)	17 (11 %)
0–0.5 h	25 (16 %)	11 (7 %)	39 (25 %)	35 (22 %)
0.5–1 h	47 (30 %)	28 (18 %)	41 (26 %)	36 (23 %)
1–3 h	55 (35 %)	70 (45 %)	48 (31 %)	46 (29 %)
3–5 h	14 (9 %)	26 (17 %)	14 (9 %)	22 (14 %)
5 h over	6 (4 %)	15 (10 %)	1 (1 %)	1 (1 %)
No answer	0 (0 %)	0 (0 %)	1 (1 %)	0 (0 %)

In contrast, Table 11.3 shows university students' daily Internet access time (excluding e-mail). On weekdays, most of them access the Internet for 0.5–3 h. There is a large difference between weekday and weekend mobile phone access time. On weekends, the university students' daily Internet access using a PC increases to 1–3 h.

Very few university students do not use the Internet, compared to approximately 20 % of high school students. The number of high school students who use only e-mail or do not use the Internet at all is greater than that of university students. University students, in contrast, use PCs, particularly on the weekend, more than high school students. This research was conducted before smartphones become common in Ehime; thus some students may shift from PCs to mobile phones after the spread of smartphones.

Internet services that high school students use most frequently with PCs are search engines, video sites, and blogs. Many of them also access search engines, blogs, and community sites with mobile phones. Many university students access search engines and video sites by using a PC, and search engines, blogs, and social networking services (SNS) using a mobile phone. As video sites require higher communication bandwidth, the PC is the best access medium. Both high school and university students' primary purposes for using the Internet are receiving, sending, and searching for information. Typical services are blogs, community sites, and SNS. This generation is nearly "digital native" and will become the largest online consumer market.

11.4   Differences in Consumer Behavior and Perceptions

11.4.1   Experience of Online Shops

Respondents' experience of using online shops (excluding Internet auctions) is shown in Table 11.4. Forty-nine percent of the high school students and 76 % of university students have used online shops. Other research shows that 82.1 % of people under-30 years (excluding students) have used online shops (IRC 2010). Compared to that research, we found slightly lower student usage of online shops. University students use online shops more than high school students. The difference is statistically significant at 1 % level. Students' low budgets constitute a strong reason for their use of online shops, suggesting that online shops can potentially become a major channel for purchases of goods and services by students.

Most student users of online shops access them via PCs, although most students generally access the Internet using both mobile phones and PCs. Only 16 % of each category of students uses mobile phones for purchasing at online shops. This finding may result from their need of a wide screen and functional usability when shopping online, which explains why PCs are students' primary online shopping device. Their non-use of public transportation may also limit their use of mobile phones outside of their house or school. The spread of smartphones, however, may spur a shift to more mobile phone use for online shopping.

Categories of goods that students have already purchased and expect to purchase from online shops are shown in Tables 11.5 and 11.6. Significant differences between number of high school students and university students are also indicated in the table. Both categories of students purchased and expected to purchase books, CDs, and clothes. University students more often purchased books, CDs, DVDs, clothes, accessories, bags, shoes, watches, and cosmetics at online shops than high school students. These differences are statistically significant at 5 % level. The goods that university students have previously purchased are similar to those that they hope to purchase. High school students, however, hope to purchase more of the

Table 11.4 Experience of online shops

Experience and devices	High school students	University students
Using online shops mainly from PC	119 (34 %)	95 (61 %)
Using online shops mainly from mobile phone	30 (9 %)	8 (5 %)
Using online shops from both PC and mobile phone	24 (7 %)	17 (11 %)
No-experience	177 (51 %)	37 (24 %)

Table 11.5 Goods that students purchased

Goods	High school students	University students	$\chi^2(1)$	P-value
Books	77	78	39.13	0.000
CDs	69	48	7.20	0.007
DVDs	43	38	11.47	0.000
Clothes	52	48	16.91	0.000
Accessories	22	22	8.17	0.004
Bags	27	26	9.06	0.003
Shoes	24	34	23.43	0.000
Watches	12	15	8.07	0.005
Miscellaneous goods	46	24	0.42	0.518
Purses/wallets	14	11	2.09	0.148
Games	44	33	6.00	0.014
Foods/drinks	8	10	5.28	0.022
Cosmetics/perfumes	14	16	7.46	0.006
Sports goods	21	12	0.48	0.488
Musical instruments/scores	22	16	2.38	0.123
Others	18	5	0.96	0.327

Table 11.6 Goods that students expect to purchase

Goods	High school students	University students	$\chi^2(1)$	P-value
Books	107	70	9.37	0.002
CDs	99	46	0.05	0.815
DVDs	82	41	0.43	0.514
Clothes	100	63	6.64	0.01
Accessories	41	27	2.81	0.094
Bags	39	32	7.68	0.006
Shoes	47	33	4.70	0.03
Watches	12	12	4.27	0.039
Miscellaneous goods	75	32	0.07	0.789
Purses/wallets	28	15	0.34	0.561
Games	66	30	0.00	0.947
Foods/drinks	30	21	2.77	0.096
Cosmetics/perfumes	26	16	1.09	0.297
Sports goods	22	10	0.00	0.971
Musical instruments/scores	38	13	0.80	0.372
Others	11	0	5.04	0.025

Table 11.7 Priorities of advantages of online shopping

Priorities of advantages of online shopping	High school students	University students	$\chi^2(1)$	P-value
Purchasing goods not sold at nearby physical shops	202 (58 %)	108 (69 %)	5.60	0.018
Shopping at home	178 (51 %)	94 (60 %)	3.54	0.060
Price/cheapness	175 (50 %)	85 (54 %)	0.74	0.389
Large variety of items	142 (41 %)	77 (49 %)	3.17	0.075
Open 24 h	122 (35 %)	77 (49 %)	9.15	0.002
Referring reviews and reputations	106 (30 %)	59 (38 %)	2.63	0.105
Purchasing at one's own pace	88 (25 %)	49 (31 %)	2.02	0.155
Delivery by other people	64 (18 %)	43 (27 %)	5.39	0.020
Easy searching of goods	92 (26 %)	38 (24 %)	0.25	0.620
Easy comparison of goods	59 (17 %)	34 (22 %)	1.67	0.197
Rich information about goods	41 (12 %)	32 (20 %)	6.61	0.010
Easy purchasing procedures	41 (12 %)	23 (15 %)	0.85	0.357
Rich information about other goods	52 (15 %)	21 (13 %)	0.19	0.660
Availability of rare goods	65 (19 %)	21 (13 %)	2.08	0.150
Speedy purchasing	23 (7 %)	18 (11 %)	3.49	0.062
Acceptable waiting time for delivery	41 (12 %)	8 (5 %)	5.44	0.020
Others	4 (1 %)	1 (1 %)	0.28	0.594

same the goods at online shops. This finding may reflect non-users' high expectations from online shops.

11.4.2 Advantages and Disadvantages of Online Shopping

Understanding users' and potential users' perceptions of the advantages and disadvantages of online shopping will contribute to online shopping market development. Table 11.7 shows students' priorities of the advantages of online shopping. Fifty-eight percent of high school students and 69 % of university students selected "purchasing goods not sold at nearby physical shops", which makes it the highest-priority feature. Approximately half the students selected as the second and third highest priority features, "shopping at home" and "price/cheapness," respectively. The fourth, fifth, and sixth priority features were, respectively, "Large variety of items," "open 24 h," and "referring reviews and reputations." University students gave more priority to "Purchasing goods not sold at nearby physical shops," "open 24 h," "Delivery by other people," and "Rich information about goods" than high school students, and high school students gave more priority to "Acceptable waiting time for delivery" than university students. These differences are statistically significant at 5 % level.

The results imply that students can largely obtain the goods that they want to purchase locally, but would shop online if the products are unavailable from nearby physical shops. Although cheaper pricing is usually a primary reason for people

Table 11.8 Unattractiveness of the online shops

Unattractiveness and anxieties of the online shops	High school students	University students	$\chi^2(1)$	P-value
We cannot confirm the goods' quality or details before purchasing	254 (73 %)	124 (79 %)	2.35	0.126
Risk of difference between goods' online appearance and that of actual goods	169 (48 %)	90 (57 %)	3.54	0.06
Risk of receiving inferior goods	195 (56 %)	64 (41 %)	9.69	0.002
Annoying advertisements after purchasing	82 (23 %)	50 (32 %)	3.99	0.046
Risk of personal information theft	106 (30 %)	46 (29 %)	0.05	0.823
Risk of nondelivery of purchased goods	92 (26 %)	33 (21 %)	1.62	0.203
Difficulty in canceling purchase orders	55 (16 %)	30 (19 %)	0.89	0.344
Too long from purchase to delivery	65 (19 %)	21 (13 %)	2.08	0.15
Complex purchasing procedure	60 (17 %)	19 (12 %)	2.09	0.148
No supply of goods in stock at many online shops	74 (21 %)	18 (11 %)	6.83	0.009
Too many items to recognize what I want	24 (7 %)	13 (8 %)	0.32	0.569
Others	9 (3 %)	1 (1 %)	2.1	0.148

shopping online, the student market clearly prioritizes availability, followed by convenience, with price only third on their list of perceived advantages. The goods sold in rural physical shops often offer less variety than those in urban shops, although physical shops are the primary channels of purchase for students. Thus, they may regard online shops as a complement of the physical shops. Therefore, small and micro online shops do not need to engage in price competition; instead, they can focus on product differentiation.

Table 11.8 displays the students' perceived disadvantages and anxieties when shopping online. Seventy three percent of high school students and 79 % of university students selected "We cannot confirm goods' quality or details before purchasing," the highest ranked disadvantage, which overshadowed the students' second- and third- ranked negative perceptions. Fifty six percent of the high school students selected "Risk of receiving inferior goods" and 48 % of them selected "Risk of difference between goods' online appearance and that of actual goods." These were elements of the second and third ranked disadvantage. Fifty seven percent of university students selected "Risk of difference between goods' online appearance and that of actual goods," as the primary element of the second ranked disadvantage. High school students gave a high priority to "Risk of receiving inferior goods" and "No supply of goods in stock at many online shops" than university students. These differences are statistically significant at 1 % level. And university students gave a high priority to "Annoying advertisements after purchasing" than high school students. This difference is statistically significant at 5 % level.

Because more than 70 % of the students selected "We cannot confirm goods' quality or details before purchasing" as the primary disadvantage of online shopping, online shops face a tremendous obstacle. Overcoming this fundamental characteristic of online shipping requires ingenuity, and technical solutions to the problem may include increasing the amount of information about product specifications and a large variety of product pictures from several angles, with the capability to zoom in and see small details. Alternatively, improved business models could reduce consumer anxiety about disappointment in the product after delivery. One possibility is making the return of goods easier and inexpensive or free to mitigate this perceived disadvantage. If an online shop sells clothes, for example, it can send articles of clothing in the next size smaller and larger than the one ordered to their customers with free returns within a given number of days. Such business models may ease customers' anxieties about online shopping.[1]

As compared with university students, more high school students report feeling anxious about the security of online shops. Because of their low budget, they would be severely affected by risks such as identity or credit card number theft. A percentage of high school students who have not used online shops may overestimate these risks. Technological enhancement of online shops' security systems for all consumer information is essential for expanding the online shopping market; in addition, it is important to prominently display complete and precise information about user security.

11.4.3 Students' Perceptions of Online Shops: Differences Among Experiences

Experiences of online shopping may strongly affect students' attitudes toward and perceptions of online shops. Some students may increase their appreciation of online shops after using them, whereas others may decrease their evaluation of online shops. Here, we examine the relationship between experiences of online shopping and students' perceptions.

Students who have used online shops indicate all factors as priorities of online shops more than those who have never shopped online. For each priority, such as "Rich information about goods," "Shopping at home," "Purchasing at one's own pace," "Delivery by other people," and "Availability of rare goods," differences between experienced and inexperienced students are not statistically significant at 5 % level. Other priorities' differences in experience ratings are statistically significant at 5 % level.

These results demonstrate that experience in online shopping improves students' perceptions of online shops. Students did not know the advantages or had only heard of the reputations of online shops before using them. After experiencing online shopping, however, students have positive perceptions of online shops, because they enjoy the many functional advantages of online shops.

We found interesting characteristics of the perceived disadvantages of online shops. Table 11.9 shows the relationship between students' perceived disadvantage of online shops and their experiences. In case of disadvantages such as "Risk of personal information theft," "Risk of difference between goods' online appearance and that of

---

[1] We should mention that this approach would be prohibitively costly and problematic from the merchants' perspective.

Table 11.9 Differences among experiences of online shops

Unattractiveness and anxieties of the online shops	Experienced	Inexperienced	$\chi^2(1)$	P-value
We cannot confirm the goods' quality or details before purchasing	210 (72 %)	168 (79 %)	3.04	0.081
Risk of difference between goods' online appearance and that of actual goods	136 (46 %)	123 (57 %)	6.05	0.014
Risk of receiving inferior goods	129 (44 %)	130 (61 %)	13.84	0.000
Annoying advertisements after purchasing	83 (28 %)	49 (23 %)	1.89	0.169
Risk of personal information theft	70 (24 %)	82 (38 %)	12.26	0.000
Risk of nondelivery of purchased goods	46 (16 %)	79 (37 %)	29.97	0.000
Difficulty in canceling purchase orders	53 (18 %)	32 (15 %)	0.87	0.351
Too long from purchase to delivery	58 (20 %)	28 (13 %)	3.95	0.047
Complex purchasing procedure	25 (9 %)	54 (25 %)	26.22	0.000
No supply of goods in stock at many online shops	73 (25 %)	19 (9 %)	21.41	0.000
Too many items to recognize what I want	17 (6 %)	20 (9 %)	2.3	0.130
Others	11 (4 %)	7 (3 %)	0.08	0.772

actual goods," "Risk of receiving inferior goods," "Risk of nondelivery of purchased goods," and "Complex purchasing procedure," students who have never used online shops rank these disadvantages more highly than those who have used them, and these differences are statistically significant at 5 % level. In contrast, students who have used online shops rank the disadvantages of "Too long from purchase to delivery" and "No supply of goods in stock at many online shops" more highly than those who have never used them, and these differences are statistically significant.

These results indicate that students who have never used online shops will often overestimate risks of online shopping. These students are aware of such technical risks, but they may not understand them sufficiently; hence, they may avoid using online shops because of their incorrect risk estimation. Other generations may overestimate the risks more because they are less familiar with ICT than such students. To expand the online shop market, we should inform users of their actual risks and show them simple methods of avoiding these risks.

Students who have used online shops indicate that the inconveniences of online shops include extremely complex user interfaces or poor distribution systems. To stimulate the use of online shops, we should develop more convenient systems in these two areas.

11.5   Conjoint Analysis

11.5.1   Model

In this study, conjoint analysis was conducted via an experiment that presented respondents with choices from which they must had to select the most desirable one by ranking their preferences. The concept of a random utility model underlies this approach. When a certain respondent n chooses item i, utility U_{ni} is denoted as follows.

$$U_{ni} = V_{ni} + \varepsilon_{ni} = \sum_{m=1}^{M} \beta^m x_{ni}^m + \varepsilon_{ni} \tag{11.1}$$

V_{ni} is an observable definite term, and ε_{ni} is an unobservable error term for a researcher. x_{ni}^m is an attribute that constitutes a choice. β^m is the marginal utility of the attribute, and this value is estimated as the sum of true marginal utility and the product of the scale parameter proportional to the reciprocal of variance of an error term. The scale parameter was normalized to 1.[2]

The probability that a respondent n will choose i from a certain choice set $C = \{1, 2, \cdots, J\}$ is denoted as P_{ni}. This behavior when a respondent n chooses i indicates that one choice's utility is higher than that of other choices j ($j \in C, i \neq j$); therefore, P_{ni} can be shown as follows.

$$\begin{aligned} P_{ni} &= Pr[U_{ni} > U_{nj}, i, j \in C, i \neq j] \\ &= Pr[V_{ni} - V_{nj} > \varepsilon_{nj} - \varepsilon_{ni}, i, j \in C, i \neq j] \end{aligned} \tag{11.2}$$

Here, if we assume that error terms ε_{ni} and ε_{nj} have the type I extreme value distribution, then the difference between the error terms has the logistic distribution.[3] Therefore, the probability P_{ni} has the following conditional logit model: CL.[4]

---

[2] This assumption indicates that the variance of an error term is constant.

[3] Refer to McFadden (1974) and Train (2009).

[4] The likelihood function of (11.3) is

$$L = \prod_{n=1}^{N} P_{ni},$$

and the log-likelihood function of this equation is

$$\ln L = \sum_{n=1}^{N} \ln P_{ni}.$$

δ_n^i is a dummy variable that assigns 1 when a respondent chooses i, and the parameter vector β can be derived as a solution of the following maximum problem.

$$\ln L = \sum_{n=1}^{N} \sum_{i \in C} \delta_n^i \ln P_{ni}$$

Table 11.10 Variety and level of an attribute

Attribute	Level 1	Level 2	Level 3
Store	Yes	No	–
Saturation	High	Low	–
Procedure	Easy	Difficult	–
Postage	Free	500 yen	–
Price	3,000 yen	4,000 yen	5,000 yen

$$P_{ni} = \frac{e^{\mu \cdot V_{ni}}}{\sum_{j \in C} e^{\mu \cdot V_{nj}}} \tag{11.3}$$

However, CL is the model achieved under two assumptions, *homogeneous preference and Independence from Irrelevant Alternatives: IIA*. Although CL is easy to analyze, it has the problem of weak model interpretability. Revelt and Train (1998) advocated the mixed logit model ML, which eliminates these two assumptions. ML is a model that has a preference from which an individual differs. When a certain respondent n chooses item i, utility is set to U_{ni}, denoted as follows.

$$U_{ni} = V_{ni} + \varepsilon_{ni} = \sum_{m=1}^{M} \beta_n^m x_{ni}^m + \varepsilon_{ni} \tag{11.4}$$

It is assumed that ε_{ni} has an independent and identical type I extreme value distribution, and the probability that the respondent n will choose i is formulized as follows.

$$P_{ni} = \int \prod_{t_1}^{T} \frac{exp(V_i)}{\sum_{j=1}^{J} exp(V_j)} f(\beta|\Omega)d\beta \tag{11.5}$$

T indicates the number of occurrences of the choice experiment, and several repetitive questions are presented to the same respondent in the usual choice experiment. f is the probability density function of β, and Ω indicates parameters such as the average and variance of β.

In this study, the choice experiment was conducted using an orthogonal array design; the variety of attributes and the level were set up as shown in Table 11.10.

Eight profiles were created using the orthogonal array design from the level of each attribute. Two profiles were combined at random, and a choice set with the added option "Using neither online shop" was created. Each respondent answered eight choice sets per questionnaire. An example of a choice set is shown in Table 11.11.

An alternative specific constant (ASC) was added to the analysis; ASC3 was introduced into "Using neither online shop." It can be interpreted as negative for the purchase of goods from an online shop if the ASC is significantly estimated as

Table 11.11 Example of choice set

Attribute	Online shop A	Online shop B	Using neither online shop
Store	No	No	-
Saturation	Low	High	
Procedure	Easy	Easy	
Postage	Free	500 yen	
Price	3,000 yen	4,000 yen	

negative. In contrast, it is affirmative for the online purchase if the value is significantly estimated as positive.

Response data of 157 students from the Ehime University and 352 students from the Ehime University Senior High School were analyzed. First, the dummy variable university dummy, whose value is 1 for a college student, was added and the cross term was multiplied by the university dummy; then, an attribute variable was added. An important difference in the attribute between university and high school students was analyzed by the time duration of online shopping. Second, the usage dummy, whose value is 1 for those who have used online shops in the past, was added and the cross term was multiplied by the usage dummy; then, an attribute variable was added, and the differences in the preference by use of an online shop was investigated.

The coefficient estimated in conjoint analysis indicates marginal utility, i.e., the increment of utility. In analysis, all the attribute variables were calculated as a random parameter, and the standard deviation parameters of the coefficient that were not significant were re-calculated as a non-random parameter. NLogit 4.0 was used as the analysis software.

11.5.2 Results and Analysis

The investigation results are illustrated in the Tables 11.12 and 11.13. The bold lettering in the table indicates the nonrandom parameters.

From the complete results, ASC is estimated as significantly negative, and respondents are considered to be positive about using an online shop. A positive and significant estimated result was obtained for saturation and purchase procedure of the site. It is considered that a respondent's utility increases when the site saturation becomes high and the site's purchase procedure becomes easy. A negative and significant estimated result was obtained for the goods price and postage, that is, if the product price and postage increase, the respondent's utility decreases.

Because the value of the marginal utility of each attribute is considered to change with the student education level and usage experience, these differences are verified as described below.

Table 11.12 Estimated result of a high school and a university student

Variable	Coefficient	t-value	P-value
Random **nonrandom** parameters in utility functions			
Availability (no)	−0.048	−0.518	0.605
Availability × univ. dummy	**0.387**	**2.467**	**0.014**
Saturation (high)	2.530	13.943	0.000
Saturation × univ. dummy	**−1.085**	**−4.144**	**0.000**
Procedure (easy)	1.167	10.580	0.000
Procedure × univ. dummy	**−0.313**	**−1.734**	**0.083**
Postage	−0.00333	−10.561	0.000
Postage × univ. dummy	**0.00051**	**1.087**	**0.277**
Price	−0.00191	−21.522	0.000
Price × univ. dummy	**0.00036**	**4.141**	**0.000**
ASC3	−6.064	−20.439	0.000
Derived standard deviations of parameter distributions			
Availability (no)	0.903	9.322	0.000
Saturation (high)	1.614	10.786	0.000
Procedure (easy)	1.063	9.433	0.000
Postage	0.003	7.419	0.000
Price	0.000	4.313	0.000
ASC3	2.869	15.992	0.000
No. of obs.	4,045		
Log-likelihood	−3,299.722		

The bold lettering indicates the nonrandom parameter

Table 11.13 Estimated result of a online shop user and a nonuser

Variable	Coefficient	t-value	P-value
Random **nonrandom** parameters in utility functions			
Availability (no)	−0.070	−0.574	0.566
Availability × use dummy	**0.390**	**2.538**	**0.011**
Saturation (high)	0.3147	12.773	0.000
Saturation × use dummy	**−1.488**	**−5.521**	**0.000**
Procedure (easy)	1.479	10.155	0.000
Procedure × use dummy	**−0.799**	**−4.444**	**0.000**
Postage	−0.00322	−8.162	0.000
Postage × use dummy	**0.00016**	**0.330**	**0.742**
Price	**−0.00200**	**−20.045**	**0.000**
Price × use dummy	**0.00027**	**3.159**	**0.002**
ASC3	−5.987	−19.486	0.000
Derived standard deviations of parameter distributions			
Availability (no)	0.935	9.176	0.000
Saturation (high)	1.635	13.587	0.000
Procedure (easy)	1.182	10.669	0.000
Postage	0.003	8.175	0.000
ASC3	3.500	15.506	0.000
No. of Obs.	4,045		
Log-likelihood	−3,278.918		

The bold lettering indicates the nonrandom parameter

11.5.2.1 Difference Between High School and University Students

For availability, high school students' marginal utility is not statistically significant, that is, they do not think availability is important when they purchase goods at an online shop. In contrast, university students' marginal utility, 0.387, is statistically significant, that is, university students consider that the purchase of the goods that cannot be obtained at a nearby store is important.

The utility of the site having high saturation is higher than that of the site having low saturation for high school and university students, but the evaluations of site saturation differ significantly. Utility for high school students is 2.530, which is higher than that of university students by 1.085. That is, when purchasing goods at an online shop, high school students attach greater importance to site saturation than university students. Similarly, high school and university students prefer to purchase at sites where the procedure is easy; however, no statistically significant difference was found between the two categories of students.

The marginal utility of the postage and product price for both students is negative, and when postage or a product's price rises, utility decreases. Furthermore, disutility from the 1-yen rise in postage is larger than that in price, that is, respondents dislike the rise of postage by even 1 yen. Although there is no statistically significant difference between high school and university students regarding postage, there is a statistically significant difference regarding product price; the marginal disutility of high school students for the 1-yen increase of product price is larger than that of university students, and high school students prefer sites with lower product prices.

The coefficient of standard deviation shows that diversity is significant for high school students. Although high school students prefer sites with high saturation, easy buying procedures, and lower postage and product prices, the degree and manner thinking vary among individuals. Furthermore, the coefficient of variation obtained by dividing the coefficient of the standard deviation of an attribute by the coefficient of the random parameter shows that the most diverse ideas existed for purchase procedure. Because all were estimated as a nonrandom parameter for the dummy variable, university students are considered to have the same preference for these attributes.

11.5.2.2 Difference in Use Experience of Online Shops

Individuals who have never shopped online consider site saturation as the most important attribute. Although experienced online shoppers also consider site saturation as important, its value of marginal utility for individuals who have never shopped online is 3.147, compared to 1.659 for experienced online shoppers (1.488 lower). A nonuser's marginal utility for ease of purchase procedure is 1.479, which is higher than that of a user by 0.799.

For postage, marginal utility is negative and no statistically significant difference by usage experience was observed. Similarly, although the marginal utility of product price is negative, the value is larger for nonusers than users, with nonusers tending to more strongly dislike increases in product price.

The marginal utility of availability for nonusers is not statistically significant, whereas it is significant for users, and they consider as important the ability to purchase goods unavailable at a nearby store.

The diversity of nonuser's preferences was observed for availability, saturation, purchase procedure, and postage, with the largest variation in the postage preference.

Although high school and university students exhibit differences in their rates of online shop utilization, it is considered that the rate generally increases with age. In this analysis, the differences between high school and university students seem attributable to their different experiences in using online shops.

11.6 Discussion and Conclusions

In this study, we examined young people's behavior and perceptions regarding online shops. Our research suggests certain important features that will support the maturation of the online shop market.

Most of our respondents use the Internet via a PC or mobile phone every day, and over half of them have purchased some goods online. The largest categories of goods that they have purchased or intend to purchase are books, CDs, and clothes. This finding indicates that online shops have become a popular channel for student purchases of goods and services. As this group represents largest future customer base for online shops, they should devise management and marketing strategies that increase students' motivation to use their products and services.

Access to goods unavailable in nearby physical shops attracts students to online shops. Surprisingly, although students typically have limited funds, they do not prioritize low prices as highly as the differentiation of goods at online shops. Students consider the most unattractive feature of online shops to be the difficulty in confirming goods. These results suggest that primary buying channel of students is nearby physical shops, and they regard online shops as a complementary option. Thus, online shops should select and market their goods or improve their business models for optimal competition with the goods and business models of physical shops.

Regarding the security of online shopping, high school students may overestimate this risk. Their small budget and the large proportion of nonusers may affect this overestimation in our findings. However, online shops are well advised to develop their security technology and prominently display their security policy. Their purchase return policy should also be prominently displayed for the student market.

Conjoint analysis clearly revealed which attribute respondents consider as important when purchasing goods online. It also shows that students consider being positive about using online shops and online shops become common for students. Although the results indicated that availability was the most important attribute in the basic research, conjoint analysis revealed that university students and Internet users highly value availability, whereas high school students and Internet users and nonusers do not consider availability as highly important.

Comparison of high school and university students as well as Internet users and nonusers revealed that the saturation of a site was considered as most important by high school students and nonusers. Similarly, ease of purchase procedure is considered to be more important by Internet users than nonusers, but no statistically significant difference was found between high school and university students' ratings of procedure's importance.

The analysis also verified that the marginal utility of a 1-yen increase differs between the postage and price of goods. The disutility of a 1-yen increase in postage is larger than that for a 1-yen increase in a product's price, suggesting that consumer behavior changes with an online shop's pricing tactics.

Because students regard online shops as a complement of the nearby physical shops, the environment affects their behavior and perceptions, even for online shops. Similar future research can provide more precise investigation of factors relating to consumers' geographical locations (urban vs. rural) and age groups.

Consumers' geographical locations may lead following possibilities. Most people who live in urban area usually use public transformations to get to their school or their office. They spend one or two hours in the train or the bus everyday. Many of them use ICT devices during the "free time" in the vehicle. In many cases, public transportation system in rural area is less convenient than urban area. Most people who live in rural area commute by their own car, motorcycle, or bicycle, and they have less time in the vehicle. Such differences of living environments may lead different usage of ICT devices and online shops.

Students, working people or house makers may use online shops in different way. Each consumer has different budget or consumer behavior. Their experiences or occupations will have close relationship with their behavior for the online security. Credit card ownership may play a number role in utilization of online shops. Consumer behavior and perceptions of online shops may be different between different countries.

Rapid spread of smartphones and tablet terminals will promote the convenience of online shops. Improving the speed of mobile communication network may improve the usability of online services. Future research will provide these investigations.

References

IRC, The state of online shopping in Ehime Pref., Monthly report of IRC, Oct., 2010
McFadden D (1974) Conditional logit analysis of qualitative choice behavior. In: Zarembka P (ed) Frontiers in econometrics. Academic Press, New York, pp 105–142

Ministry of Economy, Trade and Industry (2011) Results of Research in IT Economy Society in Japan 2010: E-commerce market survey, Feb. 2011

Ministry of Internal Affairs and Communications, Japan (2011a) Survey on ICT utilization in 2010, May 2011

Ministry of Internal Affairs and Communications, Japan (2011b) White paper: information and communications in Japan. GYOSEI

Nikkei Inc. (2010) 43th the research of Japanese retail trade. Nikkei Marketing Journal, 6.30(5)

Revelt D, Train K (1998) Mixed logit with repeated choices: households' choices of applicance efficiency level. Rev Econ Stat 80(4):647–657

The Japan Direct Marketing Association (2010) The report of direct marketing on the internet: survey of consumer's use, Dec. 2010

Train K (2009) Discrete choice methods with simulation. Cambridge University Press, Cambridge, MA

Chapter 12
Exchange of Information and Values Taking Privacy into Consideration

Hirotsugu Kinoshita

12.1 Introduction

We circulate information resources, e.g. knowledge, writing, and personal information, in contemporary society through networks with various information technology tools. Information resources in various societies need to be circulated among communities that have different values and public entities that do not belong to any particular community. Figure 12.1 outlines various relationships in communities. Members are bound by agreements based on confidence and exchanges of resources and money such as electronic money (Kinoshita et al. 2011) and local currencies. Appropriate values for the information resources and services should be evaluated before these values and information resources are exchanged in order to circulate the information resources more smoothly. Furthermore, information resources should be prevented from leaking. The goal of our approach is outlined in Fig. 12.2. The four principal components of the system consist of a copyright management system (Kinoshita and Morizumi 2008), a settlement system reflecting various values (Morizumi et al. 2010; Kinoshita et al. 2011), a secure retrieval service on Web applications (Morizumi et al. 2009), and a knowledge mining system. The copyright management system is a primitive element of our system and it enables information and values to be circulated and privacy to be protected. Section 12.2 describes a method of communication between an information distributor and an agent accompanying the information over an anonymous channel. Moreover, an anonymous-routing method for locating the position of distributed data has been proposed based on traces of routing history. By applying this method to the system that treats private information, owners of information can control the purpose of content use and protect the integrity of information (Kinoshita and Morizumi 2008). Section 12.3 explains how

H. Kinoshita (✉)
Kanagawa University, Yokohama, Japan
e-mail: kino@kanagawa-u.ac.jp

S. Uesugi (ed.), *IT Enabled Services*,
DOI 10.1007/978-3-7091-1425-4_12, © Springer-Verlag Wien 2013

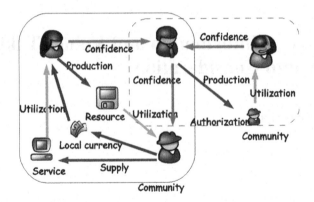

Fig. 12.1 Relationships in communities

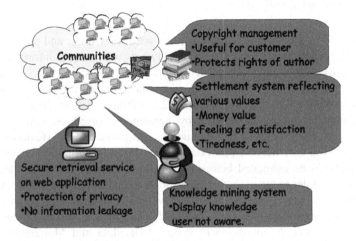

Fig. 12.2 Goal of our system

the information capsule for copyright management could be used for securitization and its concurrency. We describe the method of securitization with the information capsule, which includes mobile agents (Kinoshita et al. 2009). An investor could check the uncertainness of his/her own securities by applying this method to the system, e.g., which obligation is contained or how to estimate the value of the securities. Section 12.4 describes a value exchange system with agents that more smoothly offers information resources and services. We also created a value vector to reflect various values. We then developed a method of converting parts of vectors into securities to be circulated among the community. Furthermore, an information capsule is introduced to exchange services and local currencies (Morizumi et al. 2010).

12.2   A Network for Copyright Management and Control of Private Information

12.2.1   Background

Since the quality of digital content does not deteriorate with repeated copies, copyright protection is an important issue. To handle this issue, digital rights management (DRM) has been applied to securely deliver the contents and create secondary content quoted partially from various sources. However, DRM has not effectively dealt with an arbitration process. Moreover, conventional technologies are inflexible to control the use of copyright materials. In some cases, access to reference data in the content or data transfer is restricted when they are permitted under the copyright law. This could deprive users of the right to access. The distribution infrastructure through license agreement has improved in terms of checking copyright and licensing materials based on DRM. However, many license restrictions still make no allowance for any changes.

The system structures we are looking at are: (a) database management agent with the extended Dublin Core and access control list, (b) contents capsules containing software agents that control copyrights and access authorization, and (c) user agents. Copyright material is stored in the database. We apply the existing access control technologies to the database management agent in order to read and write data to the database (Yamada et al. 2007).

This system is expected to increase the convenience of content access. Furthermore, introducing the agent technology allows a negotiation to take place between capsule agent and access control agent for users. This will bring various advantages: for example, reducing network traffic; improving response; inspecting the sufficiency of rights; developing an efficient inference mechanism; and mediating complicated rights issues in n-dimensions.

On the other hand, protection and control of private information are serious problems in the IT enabled networks. We can also apply these methods to treat private information because access control mechanisms are efficient to protect the privacy of information. However, one of the problems of these systems is routing mechanism. Moreover, users sometimes want to use contents anonymously. On the other hand, an owner and a distributor of contents or private information want to check its integrity, update the information, and change the access rights. Because of these reasons, they communicate with each other anonymously.

In this chapter, we describe the network which is suitable for the distribution of content with DRM and for the distribution of private information controlled by the involved parties. In addition, an anonymous routing method with a mechanism to find the position of distributed data has been proposed.

Fig. 12.3 Information
capsule

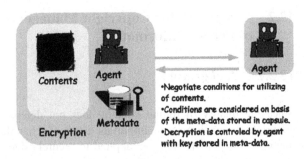

12.3 Information Capsule for the Copyright Management

The information capsule, shown in Fig. 12.3, is a framework that circulates digital contents such as music, movies, and books (Kinoshita and Morizumi 2008). It can be used to control the access to the contained information and negotiate the content usage conditions with other agents.

To control and protect the private information, the following functions are required in the system.

1. The person related to the private information can restrict the purpose of information usage.
2. The person can check the validity of this purpose.
3. The person can check the integrity of distributed information.
4. The person can change the purpose of the use of distributed information.
5. The person can update or delete the distributed information.

In particular, we require the DRM system to function as follows.

1. The owner who holds the copyright can set the conditions for information usage.
2. The identity of distributed information should be assured.
3. The owner can check the validity of information usage at the user's site.
4. The owner can change the conditions for the use of information
5. The owner can update the distributed information.

Each function of the DRM system can also be used for the function of private information control system. Therefore, we can install the private information control system on the DRM system.

The additional conditions are required to manage the secondary use of contents.

1. If a user wish to be a secondary author of contents in the secondary use, the user can negotiate with the original authors. Then copyrights are resolved between original authors and secondary author based on recorded policies with the contents. Next, new copyrights policies are recorded.
2. If the original contents are secondary contents the processes should be executed recursively.

The definition of the information capsule system structures we are looking at are:

3. The entities are the information capsule agent which manage the contents, the access control agent which processes the contents for the user and the database management agent at the author who create the contents.
4. The content are encrypted with the contents key issued by the database management agent.
5. Database management agent with the extended Dublin Core (Yamada et al. 2007) and access control list.
6. Content capsules containing software agents that control copyrights and access authorization. The copyright material is stored in the database.

Furthermore, an anonymous communication between the owner and the user of the contents should be supported if the entities needs anonymity. So the anonymous routing system (Mambo et al. 1991; Kinoshita and Morizumi 2008) is required to keep the privacy of entities.

Figure 12.3 shows the interaction of agents. An information capsule has several components shown as follows.

- Encrypted Contents: The contents which encrypted with contents key managed by the information capsule agent.
- Access Control Lists: The conditions for use of contents based on the extended Dublin Core are stored and the structure of contents which shows the components of the contents.
- Information Capsule Agent : These multiple agents are contained because each original contents has a information capsule agent.

12.4   Control of Distributed Information

The information capsule is used for DRM system (Nobukazu et al. 2003). We developed the concept of the information capsule to resolve complicated copyrights and treat the private information. The information is encapsulated with metadata and a mobile agent. The metadata includes routing and access control information. The functions of agent are to resolve rights in information and to operate the access control based on the rights.

12.4.1   Resolving Rights

The agent communicates with users and arbitrates the rights. Firstly, the agent tries to arbitrate with the metadata. Next, if the agent cannot arbitrate independently, it must communicate with a distributor based on the routing information. For example, if the content is a second use consisting of multiple sources, each copyright of source will be resolved.

12.4.2 Updating Contents

When the distributor wants to update the contents or change the access control condition, it tries to communicate with the agent. Then the agent updates the contents or meta information

12.5 Routing Based on Traces of Contents

12.5.1 Requirements of Routing

Generally, packets are forwarded based on IP address or FQDN and each packet contains a source address and a destination address. Though many anonymous networks have been found (Clarke 1999; Mambo et al. 1991). These system are assumed that all entities on network are static. Therefore, we have to consider the dynamic-moving entities because the information will be moved and distributed over P2P network. The flood routing is a solution to this condition. However, it causes heavy traffic and is not realistic. Another solution is to register the forwarding history in routers. However, this solution has two problems. First, plenty of memory is required on the routers. Second, some ID numbers are stored and this may violate the privacy.

12.5.2 Assuming Network

Another method proposed that network is composed by nodes shown in Fig. 12.4.

A capsule contains the contents, the agent that manages the access control, and routing identifiers. The functions of these nodes are to create, to use, or to forward the capsule.

The entities of assuming network are as follows.

- *Distributor* has copyrights or is an owner of distributed contents. The owner can change the access rights and purpose of contents. This function is important for the controls of digital right management and privacy.
- *User* uses the contents and creates the second use contents. Therefore, *User* may become a *Distributor*.
- *Agent* accompanies the contents and controls the content usage at the node of the *User*.
- *Router* forwards contents with meta information. This information consists of *Agent* and *Trace set* which is used for routing.
- *Capsule* is a container of the contents, *Agent* and *Trace set*.
- *Uplink message* is data or message transmitted from *Distributor* to *Agent*. This message is used for updating the contents or for controlling access rights.

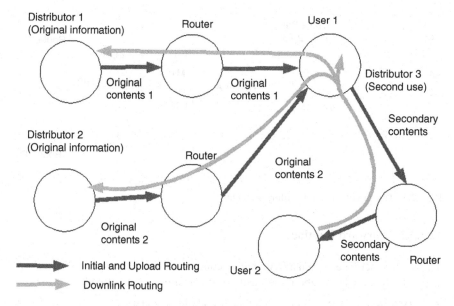

Fig. 12.4 Secondary use and anonymous communication

- *Downlink message* is data or message transmitted from *Agent* to *Distributor*. This message is used for checking the content certification when *Agent* cannot independently certify it.

12.5.3   Basic Concept

We introduce the trace matrix. It is the n dimensions matrix M_1 for each port of I. Each dimension A_i has 2^m elements. The content occupies several points (l points) in the matrix. The content has unique identification number that consists of $l \times m \times n$ bits. An example of the trace matrix with three dimensions is shown in Fig. 12.5.

   These traces will be checked when the connection between *Distributor* and *Agent* is established. Next, if *Router* finds the trace of contents, the next hop is selected from the ports which have the trace.

12.5.3.1   Initial Routing

This stage is used for the distribution and the transfer of the contents.

1. A *Distributor* or a *User* sends *Capsule* to another *User*.
2. A *Router* marks up the trace which represented as points on matrix M_1. These points are called as *Trace set*.
3. The *Capsule* is forwarded based on the conventional method with IP address or some ID numbers.

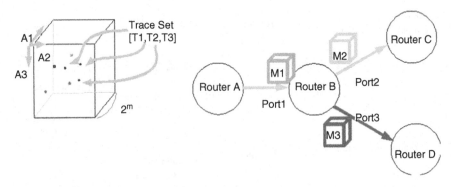

Fig. 12.5 Trace matrix and forwarding on router

12.5.3.2 Forward Routing

This stage is activated when a *Distributor* requests to contact the *Agent* of a *Capsule* without any knowledge about the position of *Capsule* on the network.

1. A *Distributor* sends a *Uplink message* to neighbor nodes.
2. *Trace set* is checked on each port of the *router*. If the traces are exist, *Uplink message* is forwarded to this port.

Figure 12.5 shows an example of routing. On *Router B*, *Uplink message* (or *Downlink message*) is forwarded from *Router A* via port1 to *Router C* via port2, if *Trace set* in *Uplink message* matches to trace matrix M_1 and M_2 and not match to matrix M_3.

12.5.3.3 Backward Routing

This stage is activated when the *Agent* requests to contact to *Distributor* with anonymous routing.

1. *Agent* sends *Message* to neighbor *Router*.
2. On each port of the *Router*, *Trace set* is checked and if the traces are exist, *Downlink message* is forwarded to this port.

12.5.3.4 Combining Multiple Rights

If the multiple contents are combined as a secondary use, each *Trace set* of the original contents and a *Trace set* of the secondary used contents are concatenated. In the stage of forwarding routing, only *Trace set* of the secondary used is checked on the *Router*. On the other hand, all *Trace set* are checked in the backward routing stage.

12.5.3.5 Resolving Multiple Copyrights

If contents contain multiple copyrights or private information and if the content usage requires some negotiation between *Distributor* and *Agent*, the *Capsule* is copied and is forwarded based on each *Trace set* such as the multicast routing. An example shown in Fig. 12.4. In this case, a secondary contents is composed from original contents 1 and original contents 2 by user 2. User1 is plays as a user, a distributor and a router. If user 2 should resolve the rights with original distributors, *Downlink message* is distributed to user 2 as a second use distributor, distributor 1 and distributor 3.

12.6 Financial Securitization with Digital Rights Management System

12.6.1 Background

Subprime mortgage crisis is a serious financial problem (The Rise and Fall of Subprime Mortgages 2007). The financial securities of the subprime lending are issued by using the structured finance. Because of complexities of the structured finance (The role of ratings in structured finance: issues and implications 2005), the securitization of the assets, distribution of the securities and the protection of the private information among transactions of secirities on IT enable networks became serious problems. Especially, uncertainness of the securities that which debts are included the securities. On the other hand, we studied an application for the privacy information control using the digital rights management system (Kinoshita and Morizumi 2008). In this paper, we show that the information capsule for the copyright management could be used to the securitization and its concurrency. We propose the method of the securitization with the information capsule (Kinoshita and Morizumi 2008; Yoshioka et al. 2003) which includes mobile agents. By applying this method to the system that a investor could check the uncertainness of the own securities. For example which obligation is contained or how to estimate the value of the securities.

12.6.2 Problems on the Securitization

The securitization is a structured finance process (The role of ratings in structured finance: issues and implications 2005). The cash flow from loan, obligation or mortgage backed security of the debtors producing financial assets are securitized.

Figure 12.6 shows the entities who appeared in securitization. The originator initially owns the assets engaged in the deal. Special purpose vehicle (SPV) is a legal entity. An originator negotiates the assets to the SPV to manage the assets and

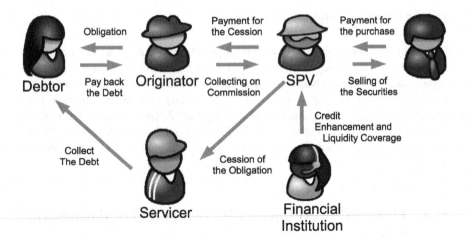

Fig. 12.6 Securitization

avoid the risk. Then SPV issues securities and sells these to the investors. A servicer collects payments and monitors the assets. The servicer can often be the originator. A financial institution reinforce the credit enhancement and the liquidity coverage.

Generally the securities are pooling and repackaging into another securities. Pooling is a resource management term that refers to the grouping together of resources for the purposes of maximizing advantage and/or minimizing risk to the users. Securities have been issued to repackage existing assets which makes them more attractive to investors.

Tranching is the method that the securities are divided into categories as part of the same transaction. Each category has same level of the risk. The more senior rated tranches generally have higher ratings than the lower rated tranches. For example there are rated the class AAA, class AA, class A, class B. The tranching is an important mechanism of the structured finance. Generally, they are paid sequentially from the most senior to most subordinate.

Dividing is the process that the securities divided into a small amount of bonds.

The repetition of pooling or repackaging and tranching or dividing increases complexity that which debt is included in the securities and the influence of the debt.

12.6.3 Securitization and Rights Management

To control the processes of securitization, the following functions are required in the system. Especially the function of the secondary use plays an important role.

1. Debts such as loans could be treated as the contents.
2. An investor can recognize that which securities are contained in the securities.

3. An investor who has the securities can check the results of the investment from each security.
4. The change of status of the cash flows are propagated from the servicer to the investors.
5. The debtors can check the negotiation of the securities.

Each function of the DRM system can also be used for the function of the securitization system. Function 2 and 3 are installed as the function of resolving copyrights on the DRM system. Function 4 is installed as the function of updating contents. Function 5 is installed as the function of check the purpose of usage. Therefore, we can install these functions for the securitization on the DRM system.

12.6.4   Processes of the Securitization

12.6.4.1   Securitization

In our system, the structure of the securities is defined as follows.

The information capsule of the securities contains a body of the agent, a body of the securities, a key for the encrypted securities, a identifier of the debtor or SPV, a contract described in XML, an evaluation method of the result of the investment and conditions of management in repackaging and tranching.

If the security is repackaged, one agent and information of the source or original securities. In this case, we call the agent of the repackaged security and agent of the sources the master agent and the slave agent, respectively.

First, the debtor issues a bond to the originator. The structure of the bond is similar to the securities except the issuer is the debtor. The bond is transferred from the originator to the SPV. Next the SPV issues the capsule of the securities to the investor. The copies of the securities are then sent to the servicer. The agent at the debtor and servicer exchange the information and the servicer monitors the cash flows.

12.6.4.2   Repackaging and Pooling

Figure 12.7 shows the process of the repackaging. This process is similar to the process of secondary use in copyright management. All bodies of securities are concatenated or processed appropriately as new, repacked securities. Each information capsule of securities is concatenated or aggregated. The aggregation is key to the repackaging. First, the aggregated agent inherits the functions of previous agents. Then identifiers of debtors are stored into a database.

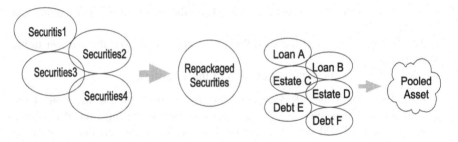

Fig. 12.7 Repackaging and pooling

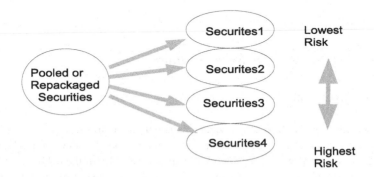

Fig. 12.8 Tranching

12.6.4.3 Tranching and Dividing

Figure 12.8 shows the process of the tranching. First, the agents are copied to tranched securities. Next, the body of securities and contracts is reorganized. The dividing process is similar to that of tranching except that the ranks have the same level of risk.

12.6.4.4 Operations of Securities

An example of the currency of the securities is shown in Fig. 12.9.

The originator pools several of the debtors' loans. Next, the pooled loans are transferred to the SPV. The SPV then issues the securities on the basis of the pool. The SPV transfers the obligation to the servicer.

Another originator and SPV pool these issued securities and issue repackaged securities as new assets.

When an investor values his or her securities, the agents operate as follows. The master agent checks the dependencies of the securities and makes a tree of the

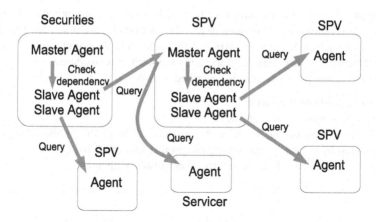

Fig. 12.9 Operating of securities

structure of the securities. Each node has a method to value the securities located at lower levels of the tree. The types of method are as follows:

- Immediately: The securities are evaluated independently.
- Query: A query to the SPV is required. If it is *Query*, the dependent slave agent is activated and obtains the information, e.g. losses of the portfolio, from the agent at the SPV.

The agent at the SPV that receives the query from the slave agent, checks the collection of the debt, and reports the results. If the securities are repackaged, the query is executed recursively.

The advantage of this scheme is that the queries from investors to servicers are distributed.

12.7   A Local Currency System Reflecting Variety of Values

12.7.1   Background

The local currency is suitable for financial settlement between members of a community. Several theoretical approaches have been studied to analyze the circulation of local currencies. Katai et al. (2009) described the ability of local currencies by introducing a method based on fuzzy network analysis. Miura (2008) explained how local currencies circulate on the basis of the search theory of money. As these previous studies treated the values as a criterion, just circulating resources in the communities is insufficient to meet our aims. Muto (2006) developed theoretical foundations of social motives in a dyad model from the viewpoint of altruism and egalitarianism. This study evaluates the values and motives in a society.

We propose a value exchange system with agents for smoother exchange of information resources and services. When the transactions are done, the possibility of the information leakage is detected through multiple communities, and the balance between convenience, safety, and circulation is considered.

Outline of the whole system is described as follows.

1. The definition of the value
2. Settlement between two entities
3. The circulation of the value with securities. The credit of the securities is evaluated by a human relationship diagram (Kubo et al. 2011).
4. The settlement based on the information capsule with agents.

12.7.2 The Exchange of Values

12.7.2.1 Values and Services

When information resources and services are supplied through a network, their values are unified and expressed in prices in conventional settlements. Otherwise, the supplier is either voluntarily giving them away or gaining advertisement through sponsorship. Furthermore, although the price is in the local currency, finding appropriate parties with which to exchange and the services for currency can be difficult. It is also difficult to exchange one local currency for services in different communities. Thus, various types of value should be considered to describe a user's conditions for the transactions, and the information capsule with the agent is required to exchange services between communities. Certain values, e.g. laws, ethics, or feelings of satisfaction, are difficult to replace with the conventional value of money. Additionally, resources that are operated independently and are used among different services need to be used without the information leaking or being falsified. Private information, e.g. personal information and transaction history, should be distributed actively to supply adaptive services. In this situation, privacy must be under the control of the subjects whose information this is.

12.7.2.2 The Implementation of the Scheme

The management of the traditional local currency is categorized into three types i.e. the bill, the bankbook, and the account.

Our scheme adopts the bill method because we think it can best maintain the privacy of the participants.

We proposed an electronic money system (Morizumi et al. 2010), a DRM system based on the information capsule (Kinoshita and Morizumi 2008), and a securitization system (Kinoshita et al. 2009).

We combined these systems to implement our local currency system.

Fig. 12.10 Classification
of the value

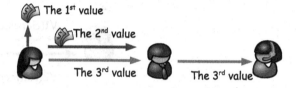

Fig. 12.11 Value vector

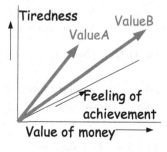

12.7.3 Circulation Using Securities

12.7.3.1 Classification of Values

We classify the values into three categories shown in Fig. 12.10.

- **The first value:** This value is effective for oneself.
- **The second value:** This value is effective for entities who transact with each other.
- **The third value:** This value is recognized commonly in the community.

It is difficult to circulate the first and second values one after another. To circulate the first and second values, we have to convert them into the third value.

12.7.3.2 Value Vector

We describe the value as the vector shown in Fig. 12.11. Each axis shows a value. Let $x_1, x_2, \cdots, x_n$ be the values. The value of object x is denoted as

$$V_x = (x_1, x_2, \cdots, x_n).$$

Services, products, and local currencies have a value vector. We introduce two types of value vector functions. One is the transaction evaluation function $F_{trans}e(V_x, V_y)$, which shows the gain of the transaction from the viewpoint of the entity e. V_x and V_y are a value vector of a service and a reward for the service, respectively. $F_{trans}e$ becomes positive if the transaction yields a profit for e. The

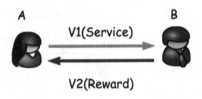

Before transaction	$V_A(t)$	$V_B(t)$
After transaction	$V_A(t+1)$	$V_B(t+1)$

Fig. 12.12 Transaction

other is the property function $F_{prop}e$, which shows amount of the property of the entity e shown in Fig. 12.12.

12.7.3.3 Transactions

The transaction comes off successfully if

$$F_{transA}(V_1, V_2) > 0 \wedge F_{transB}(V_2, V_1) > 0.$$

Let $V_A(t)$ and $V_B(t)$ be the value vectors at the time t of entities A and B respectively. Then, the amounts of property the entities have after a transaction are described as

$$V_A(t+1) = F_{propA}(V_A(t), -V_1, +V_2)$$
$$V_B(t+1) = F_{propB}(V_B(t), +V_1, -V_2)$$

12.7.3.4 Container of the Evaluation Functions

The evaluation functions $F_{trans}e$ and $F_{prop}e$ are represented by a combination of equations and look-up tables. We call this a function container. The mobile agent, a part of the information capsule, uses the container to evaluate values when the transactions are requested. Each entity has to register the evaluation functions in advance.

12.7.3.5 Conversion of the Values into Securities

The first and second values are converted into the third value by issuing securities. Some parts of values that an entity receives may be the third value and can be circulated one after another. Other parts of values may be the first or second values.

Fig. 12.13 Conversion
of the securities

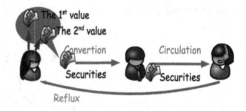

In our system, any entity can issue securities as a local currency. Let V_S and V_R be the value vectors for a service and a reward, respectively. Let $\cup$ be the conjunction of parts of the value vector. V_R could be divided into three parts shown in Fig. 12.13.

$$V_X = V_{1st} \cup V_{2nd} \cup V_{3rd}$$

where $V_{1st}, V_{2nd}, V_{3rd}$ are the value vectors of the first, second and third, respectively. For example V_{1st} is denoted as

$$V_{1st} = (x_1, x_2, \cdots, x_n)$$

If a subset V_{profit} of the value vector, which is a subset of $V_{1st} \subset V_S$, is profit for the entity, securities S_{profit} are issued instead of the first value. The value vector of S_{profit} is denoted as

$$V_{profit} = (x_1, x_2, \cdots, x_m)$$

Similarly, V_{2nd} is processed.

12.7.4   Credit of the Securities

12.7.4.1   Reputation of the Personality

Outline of the Method

The reputation of the entities consists of two parts. One is evaluated by the performance history of the securities issued by the entity. The other is the subjective reliability, which is evaluated by the relationships in the communities.

The human relation diagram is a graph that shows the relationships between entities in the community and shows the reliability of the partners in the transactions (Shimizu et al. 2008). For example, the relationships could be a friendship or a business connection. This concept can be seen in the social network services. An example of the diagram is shown in Fig. 12.14.

Fig. 12.14 Human relation
diagram

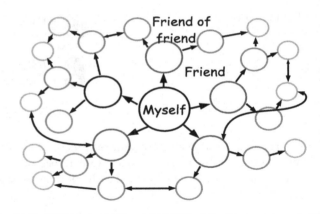

12.7.5 Currency with Information Capsule

12.7.5.1 Categories of the Information Capsule

In our system, the local currencies, the services that can be supplied through the network, and the list of suppliable services are circulated by the information capsules.

The information capsules are categorized as follows.

1. **Local currencies:** The information capsules of the local currencies includes the third values and securities converted from the first and second values.
2. **Service supply:** The service supply is used to advertise the services that the entities can supply to the community. The contents consist of the entity who wants to supply, the service, and the value vector of the service. The capsule of the service list is circulated among the entities in the community such like the super-distribution.
3. **Service demand:** The service demand is used to find out the services required in the community. The contents consist of the entity who wants the service and the value vector of the service.
4. **Service body:** The service body is categorized into four types on the basis of the location of the supplier. We assume that the supply of the products is considered as a service.

   (a) **On site services:** These services require face-to-face interaction between entities or the work on site in the real world.
   (b) **Real-world products:** These services provide real-world products. The supplier transfers these to the customer.
   (c) **On line services:** These services are provided through the network.
   (d) **Digital contents:** These services provide the products that are distributed through the network.

The contents of the information capsule for the real-world services and the real-world products are certifications of the services shown in Fig. 12.15.

Fig. 12.15 Types of services

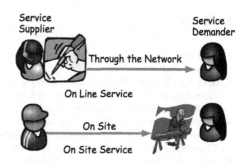

12.7.5.2 Protocols

User Agent

Each entity has a user agent to negotiate with the agent contained in the information capsule.

Finding Services

- An entity who wants the service distributes the information capsules of the service demand to the community.
- An entity who can supply the service distributes the information capsule of the service demand to the community.
- The service supply agent, the a service demand agent, and the user agent exchange the information about services.
- If supply and demand match, the value of the service is presented.

Evaluating Values

- The service supply agent and the service demand agent evaluate the value vector using the evaluation functions mentioned above.
- The service demand agent present the value vector in the local currency to be used for the payment.
- If the results of the value evaluation benefit each entity, the transaction comes off successfully.

Payment

The information capsule of the local currency is moved from the customer entity to the supplier entity.

References

FreeNet White Paper Original white paper by Ian Clarke (1999) Division of Informatics, University of Edinburgh

Katai O, Kawakami H, Shiose T (2009) Fuzzy local currency based on social network analysis for promoting community businesses. In: Gen M et al (eds) Intelligent and evolutionary systems, vol SCI 187. Springer, Berlin, pp 37–48

Kinoshita H, Morizumi T (2008) A network for copyright management and control of private information. In: The 2008 symposium on applications and the internet the workshop on ITeS: IT enabled services (ITeS), pp 448–451 (2008–7)

Kinoshita H, Morizumi T, Suzuki K (2009) Financial securitization with digital rights management system. In: International symposium on applications and the internet (IEEE/IPSJ), ITeS 2009, pp 197–200(2009–7)

Kinoshita H, Tajima Y, Kubo N, Morizumi T, Suzuki K (2011) A local currency system reflecting variety of values. In: International symposium on applications and the internet (IEEE/IPSJ), ITeS 2011, pp 562–567(2011–7)

Kubo N, Morizumi T, Suzuki K, Kinoshita H (2011) Multi agent system in order to distinguish the floating context of personalities. In: IEICE the 2011 symposium on cryptography and information security, SCIS2011, 1F1-1

Mambo M, Kinoshita H, Tsujii S (1991) Communication protocols with untraceability of sender and receiver.D-I Vol. J74-D-I No. 7, pp 429–434

Miura K (2008) A theorical analysis for the circulation of the local currencies. Hosei University Repository, ISSN0387-2610, http://hdl.handle.net/10114/1667, pp 69–76

Morizumi T, Suzuki K, Kinoshita H (2009) A system for search, access restriction, and agents in the clouds. In: International symposium on applications and the internet (IEEE/IPSJ), ITeS 2009, pp 201–204(2009–7)

Morizumi T, Kudo M, Suzuki K, Kinoshita H (2010) An electronic money system as substitute for banknotes. In: International symposium on applications and the internet (IEEE/IPSJ), ITeS 2010, pp 316–319 (2010–7)

Muto M (2006) A theory of social motives on altruism and egalitarianism. Sociol Theor Method 21 (1):63–67

Shimizu K, Toda A, Kinoshita H, Morizumi T (2008) An introduction to the evaluation of human relationship diagram in a community currency of local SNS. Technical Report of IEICE, vol 108, no 331, SITE2008-37, pp 7–12

The Rise and Fall of Subprime Mortgages (2007–11) Federal Reserve Bank of Dallas

The role of ratings in structured finance: issues and implications (2005) Bank for International Settlements (2005–1) net The Workshop on ITeS: IT enabled Services (ITeS) (Aug.2008).

Yamada K, Kinoshita H, Morizumi T, Inazumi Y (2007) Convenience improvement of using contents that used information capsule of agent base. In: Symposium on cryptography and information security, SCIS 2007, p 214 (2007–01)

Yoshioka N, Thahara Y, Honiden S (2003) Active contents: flexible contents distribution by mobile agents. IPSJ J 44(SIG19(TOD20)):45–56

Chapter 13
Real Name Social Networking Services and Risks of Digital Identity

Can We Manage Our Digital Identity?

Yohko Orito

13.1 Introduction

The development and social penetration of Information Communication Technologies (ICT) has been intergraded into various aspects of our lives. Nowadays many Internet users are enjoying the use of free web-based services such as Social Networking Services (SNS), search engines, video access,[1] to name just a few. The number of users continues to rise, and SNS which encourage users to disclose their real names on their websites seem to have boosted their internet presence even further.

It is alleged that 'real name' SNS enhances trust which leads to more frank and clear communication among website users.[2] In fact, it seems that many users consider inappropriate or impolite manners become the norm when the users do not disclose their real name. Furthermore, many users may appreciate the convenience of suggested friends or websites that occur when real name systems on the SNS websites are utilized. These trends on such SNS have been receiving growing attention on the web.

On the other hand, the social impacts of social media and SNS have been analysed in various contexts (for example, gender issues (Asai 2009, 2010), and influence on children (O'Keeffe and Clarke-Pearson 2011). One example is the concern it raises over information privacy. Because providers of SNS have conducted promotional activities to collect personal information, many SNS users already have provided huge amounts of personal information including personally identifiable information

---

[1] The recently Megaupload debacle has had a huge impact on legal ramifications.

[2] For example, the international manager of Facebook explained that Facebook enables the users to construct a reliable relationship with other users using a real name system (see, http://ascii.jp/elem/000/000/135/135542/).

Y. Orito (✉)
Faculty of Law and Letters, Ehime University, Matsuyama, Ehime, Japan
e-mail: orito@LL.ehime-u.ac.jp

S. Uesugi (ed.), *IT Enabled Services*,
DOI 10.1007/978-3-7091-1425-4_13, © Springer-Verlag Wien 2013

on these websites. There are social concerns about the protection of information privacy and online stalking. In order to allay such social concerns, many SNS have launched various privacy settings or options for their users.

However, even if various privacy options or privacy settings are available for SNS users and the users understand that they themselves also take a certain level of self responsibility to protect their information privacy, the actual responsibility, whether legal or moral, remains murky at best when the users reveal their real name on the websites. Although the users who give out their real name can enjoy information sharing with each other, they may lose complete control over their digital identity.

This "digital identity" can be defined as information to "describe the persona an individual presents across all the digital communities in which he or she is represented" (Odin Lab 2010). The social influence of SNS where the digital identity of individual users is developed remains a matter of debate. This chapter examines the social impact of real name SNS on individuals from the viewpoint of digital identity risk management. (Hereafter, "SNS" should be interpreted to mean "SNS which encourages users to reveal their real names," unless otherwise specified.)

13.2 Social Networking Services (SNS)

13.2.1 Business Model and Architecture

SNS have been attracting a growing number of users; clearly a response to the user perception that there are significant benefits to membership. For instance, SNS enable users to activate and broaden their network of "friends". On the relevant website, users can enjoy sharing information such as their profile, photos and feelings with other users, and are also able to search for and find existing friends in real life or even new friends with similar interests or the same taste on a world wide scale.

In many cases, using SNS are free and large parts of their operations are covered by advertising revenue (see Negoro 2006). This relationship is one of the root causes of the SNS debate covered in this chapter. With the purpose of getting advertising revenue, the providers of SNS websites need to make the website as attractive as possible for the advertising clients. Generally, the higher the website access rate, the higher the value of the SNS for the advertising media. To generate more traffic to the SNS website, it becomes necessary for the SNS that it increases both registration numbers and the quality of users' personal information. This situation has facilitated the current circumstance in which the business model has to balance user privacy with commercial sense.

   Providers of SNS websites work constantly on expanding the number of users and on encouraging the information sharing among the users, based on the concept of network externalities or network effects. As a result, continuous improvement of SNS information systems is absolutely vital for its ongoing operation. Critically, SNS have been designed to enhance information sharing among users. As well as the architecture of search engine systems which collect huge amounts of information individuals provide such as search key words (for example Battelle 2005), it has been considered that the more a SNS website is used by users, the more personal information it acquires about its users, and subsequently the better the quality of information users can get from the SNS website.

   This type of information system can be described as a "dataveillance system[3]" (Clark 1988), which is designed to bring about continuous improvement of the quality of information services provided to individual users (Orito 2011). On that basis, SNS information systems built on this dataveillance architecture collect user information and utilize it to attract further users on an ongoing basis. Fundamentally then, the operation of dataveillance systems is crucial for the SNS business model. When considered in conjunction with the attraction of advertising clients through target advertisements or behavioural marketing, the collection of users' personal information is indispensable.

13.2.2   Case of Facebook

Facebook is one of the most famous SNS websites on the Internet. It has gained a significant market share and is estimated to have over 700 million users around the world, including Europe, North America, and Asia.[4] Some users' communication on Facebook has been used for pro democracy perspectives, and accordingly, their impacts on various political activities have attracted interest. The most distinctive feature of Facebook is its real name system. When the user signs up on its website, the user is encouraged to give their real name on the site, and already many Japanese do so in Japan (Mobile Marketing Lab 2011). This is in contrast to Japanese SNS in which a pseudonym system is maintained. As noted earlier, it is alleged that this real name system increases trust among Facebook users. For example, Dwyer et al. (2007) demonstrates survey results that detail Facebook members expressing greater trust in both Facebook and its members, compared to members of MySpace which had adopted a pseudonym system at that time.

   Like other SNS websites, Facebook has developed dataveillance component as part of its information system. Personal information the user provides to Facebook is analyzed and utilized for the operation of Facebook websites (Weekly Toyokeizai 2012). It is assumed that the more detailed the personal

---

[3] This word is coined by Clark (1988).

[4] See, http://www.internetworldstats.com/facebook.htm.

information they provide is, such as family information, hometown, photograph of users' face, old school name, political view, religious affiliation, date of birth and so on, the better the analysis for friend recommendation and friend search. Through the mechanism, it is possible to increase the number of Facebook users, and to enhance its value from the standpoint of the advertising media. In particular, personally identifiable information is useful for encouraging users' participation on Facebook, because it is observed that this information is more closely related to the real life of the users, and users have a tendency to trust information posted by the users revealing their real names.

13.2.3 Privacy Settings and Users' Responsibility

As pointed out earlier, discussions over user privacy protection on SNS like Facebook have been central to this business model. In order to deal with widespread concerns over information privacy, almost all SNS provide privacy settings as well as other online services to allay users' fears. On a typical privacy setting, the user can define to what extent they disclose their personal information on the website. However, some argue that such privacy settings are not easy to understand (e.g. Lawler and Molluzzo 2007) and accordingly do not work well to protect information privacy. Needless to say, these settings and privacy policies need continual improvement. In the case of Facebook, users should take responsibility to disclose their personal information with the privacy options Facebook provides.

Facebook gives its users little alternative to using their actual names. Facebook states that users' accounts could be deleted if the user selects a nickname in the "Statement of Rights and Responsibilities" as follows (complete details can be found at: http://en-gb.facebook.com/legal/terms).

> 4. Registration and Account Security statement
> Facebook users provide their real names and information, and we need your help to keep it that way. Here are some commitments you make to us relating to registering and maintaining the security of your account:
> 1. You will not provide any false personal information on Facebook, or create an account for anyone other than yourself without permission.
> 2. You will not create more than one personal profile. If we disable your account, you will not create another one without our permission.
> ......
> 10. If you select a username for your account we reserve the right to remove or reclaim it if we believe appropriate (such as when a trademark owner complains about a username that does not closely relate to a user's actual name).

In fact, it has been reported that users who do not use their real names on Facebook were expelled from using Facebook without warning.[5]

---

[5] See, Facebook giveth, Facebook taketh away (http://www.smh.com.au/news/web/facebook-giveth-facebook-taketh-away/2007/10/29/1193555573838.html?page=fullpage#contentSwap1).

According to Kirkpatrick (2010, p. 199), Zuckerburg (CEO of Facebook) said "Having two identities for yourself is an example of [a] lack of integrity." However, is it possible for people to behave as a social being without resorting to different personas depending on the situation, even if it is on SNS? Also, can we effectively take complete responsibility to control our personal information and to construct the digital identity that we would want? The digital identity of SNS users seems to be problematic for the users themselves to control. In the following sections, the uncertainty of controlling personal information and digital identity is discussed.

13.3   Out of Complete Control

13.3.1   Disclosure of Information by others

As already noted, almost all SNS attempt to provide privacy options to their users so they are able to manage their information privacy while trying to balance this with a users' real name policy. However, even if SNS users can choose their privacy settings and accept the self responsibility principle, he/she cannot have complete control over the flow of personal information concerning him or her and they cannot control events that may shape their digital identity. One of the reasons is that information related to one user could be posted not only by one's self, but also by others.

Let's assume that an individual (hereafter referred to as, "A") posts a blog article which describes a humorous episode at a party, and then uploads pictures of party participants with tagged information including their names. Even if the blog is only accessible for exclusive members, that is, friends who are linked, the uploaded picture is not controlled by the participants of the party. If one of the participants (hereafter referred to as, "B") had not wanted these pictures to be posted by A, it would be likely that the picture files could not be completely eradicated, perhaps due to copies having been made despite B having asked A to delete the files. As a consequence, it may become a serious relationship problem between A and B. In reality, cases just like this have happened many times on SNS websites.[6] There is a possibility that certain information concerning one user is revealed by other users in a way that is not desirable from the first user's perspective. This type of situation may put a burden of self-responsibility on individual users. Moreover, there is an asymmetric level of impact and responsibility between users, one of which uses their real name and one who does not. For example, when one user using a handle name posts a compromising message concerning another user who uses their real name on the SNS website, it could be considered that there is a much greater impact on the user using their real name.

---

[6] See, Trend micro survey (http://jp.trendmicro.com/jp/about/news/pr/article/20120410035738. html), and New York Times (Spring Break Gets Tamer as World Watches Online, www.nytimes. com/2012/03/16/us/spring-break-gets-tamer-as-world-watches-online.html?_r=2&src=recg).

13.3.2 Interpretation to Information

The second reason why the user loses complete control of their digital identity or information concerning him or her on an SNS site is related to interpretation or context-sharing problems. The meaning of information depends on the context, but it is difficult to interpret information in the original context by others completely, even if the information sender and receiver have developed a close relationship. In particular, when certain personal information is coupled with other information or just partially referred to, meaning which is different from the original context may be ascribed to the original information by other users. It has been argued that such a distortion of information could be amplified, particularly when information processing is repeated (Murata 2004).

For example, if one user (hereafter referred to as C) misunderstood the meaning of blog articles posted by his/her friend (hereafter referred to as D) and got angry with D, C may write something derogatory about D on the SNS. In that case, even though it was only a misunderstanding that made C post such comments, the information is stored and disclosed on the website. Under such circumstances, like a celebrity, if our message, connected with our real name, is misunderstood by other users, whose responsibility would it be? This is an example of the accompanying risk when we disclose information on the Internet.

The users' list of friends also becomes the subject of the interpretation. Even if the users were to accept the self-responsibility principle and restrict the access to their page, it is difficult to take responsibility for other users' interpretation on digitalized information the original user transmits. Additionally, the interpretations of information concerning a user's activity could be ever-changing. It's not always true that information which is revealed at a certain point will be interpreted in the same context in future situations. Nobody can understand how a context could be change in the future. There is no certainty that what may be considered harmless information at present may impact negatively in the future.

13.3.3 Architectural Control

The third factor is related to architectural control. As referred to earlier, SNS could be considered as a dataveillance system which collects and utilizes certain aspects of the information that a user provides; for example, to make appropriate suggestions or recommendations for other users. When we consider this, we need to realize that even search keywords and page access information may be utilized for further recommendations to prospective friends. In other words, our facial photo tagged with our real name, could be provided to other users.

However, most users don't know how their names or photos are used, nor what kind of recommendation list the SNS might present to other users. The average user simply cannot understand why their profile might be considered suitable for recommendation to other users, because the SNS as a dataveillance system is a black box for them (Orito 2011, p. 13). Whereas developers of SNS supposedly have the intention to expand user relationships on their website, our human relationships are analysed based on SNS web architecture without users' understanding. Sometimes users may even have blind confidence in these recommendations. The SNS as a dataveillance system can make suggestions for users before the users may even consider them, and often users accept these ideas. Most users cannot recognize the architecture influence that such a situation implies. Thus there is uncertainty over how users' personal information is used by the system architecture of SNS websites.

Moreover, the architecture of SNS is designed to stimulate users' feelings to want to know what friends and friend's friends do, and feelings of wanting to attract other users' attention on the SNS. For example, on the Facebook website, many active users want to attract other users' "Like this" or comments, by submission of certain articles, photos, and so on. This mechanism means that such users' feelings can reinforce dataveillance systems which continue to collect, store, and utilize personal information the users provide, and the users could follow the principle that is describe the network externalities or attention economy (Goldhaber 1997) through the frequent use of SNS.

13.4   The Dull Mirror World

Those SNS which encourage their users to reveal their real names and to develop real social relationships with other users seem to have the intention to construct a *"mirror world"* around the users on their website. This mirror world has no geographic or time constraints. Also, the mirror world reflects users' real life to some extent; a real life that could be affected by the mirror world. In other words, the digital identity which is formed on the SNS website can affect social relationships in the real world (Rogerson 2002).

The information garnered from the user and stored on the SNS website defines how the user expresses themself in the mirror world. However, as noted in previous sections, the user information cannot be controlled in isolation by themself. Digital identities can become distorted versions of the real life identity. It is also difficult to distinguish relationships developed in the mirror world.

Thus, it is not necessarily the case that the mirror world developed on a SNS website can reflect the reality that the user may wish to project. Sometimes, a user's digital identity can be formed by others and even system architecture, intentionally or unintentionally. The personal images developed in the mirror world can diffuse rapidly across SNS websites, and such digital identity is exposed to a variety of interpretation by others. From that standpoint, the mirror world which is shaped via SNS websites could be regarded as a dull mirror, even though the users give out their real names. Also, the dull mirror world continues being recorded in one way or another. Like life logging systems (Allen 2008), it is thought that information about one's private life, continually being saved on the SNS website, may annoy the user.

13.5 Dilemma of Real Name System

There is a way of thinking that by maintaining a relationship on the SNS website, communication in both real and virtual space can be managed effectively, and that this is best done by revealing the user's real name. However, the ability to control how one's digital identity is developed is lost to some extent on an SNS. It is inevitable that others will post information related to another user indirectly or purposefully on the SNS, and that information provided by one's self and others is interpreted in differing contexts of others, such as friends or even a third person. The dataveillance systems, operating as a black box, continue to analyse which kind of information should be provided for users in a timely manner and to encourage users to reveal their personal information.

Such situations pose a dilemma for SNS that require real name users. It has been shown that while the user would typically like to continue communicating under their real name, their capacity for shaping their digital identity could be impaired to some extent. On the other hand, if SNS were to move away from using a real name policy, opportunities to receive useful and appropriate information for oneself may well decrease, and it may even negatively affect the opportunity to expand their social relations. In fact, the transmission of information through the SNS website makes it possible for the user to receive a huge amount of responses from other users. It cannot be denied that there is a real advantage at present in this regard for SNS using a real name policy. Users are forced to confront this dilemma; to risk or not risk using their actual name.

Recruiting activities are a typical example. Recently, companies in the U.S.A. provide a "job fit score" which is calculated based on users' information they reveal on various social media (Garling 2011), such as Facebook and Twitter. For the job hunter, these services may force them to reveal personal information via social media, because digital identity can be taken as a good sign of employment awareness. On the other hand, with regard to the dilemma described earlier, it is not easy to control our digital identity by ourselves.

Moreover, if we can understand such a SNS dilemma, how can users continue accessing SNS under such situations? When we develop our digital identity and personal information on the SNS, how can we keep a close watch on our information in the mirror world? It sounds like monitoring for oneself by oneself. And if a job hunter were to consider that the company that they were applying to would check their personal information stored on the SNS, should they then change their privacy setting and blog articles to build an appropriate digital identity, one the company wants? It may not necessarily be that information sharing on SNS is a utopia after all.

13.6 Social Challenges for Future SNS

It is expected that as SNS utilizing a real name policy like Facebook become more popular, more SNS will collect personal information which is connected with users' real activities. As these SNS spread out more and the personal information collected in the SNS comes to be utilized more widely, then the personal information which is stored in each SNS websites may be merged into one large database for advertising purposes.

Under such circumstances, further social concerns will be raised. One of the issues is whether a private large organization such as Facebook, can effectively take responsibility to balance its commercial interests with its social role. If private organizations cannot be responsible for the data collection of their users, and appropriate regulations cannot be implemented, digital identity in the dull mirror world may well seriously impact on our real life. In fact, one of the Face book policies is "Done is better than perfect" (Weekly Toyokeizai 2012, p. 53). If the operation of Facebook had negative impacts, can they take responsibility for the damage of them?

Another important issue is that related to the authentication system that guarantees the real name of users. Without such authentication, the accuracy of the information cannot be guaranteed. If someone made a false digital identity of a SNS user, its influence could cause an incalculable loss on the user. In order to avoid this kind of situation, measures need to be taken to protect the individual's digital identity.

13.7 Conclusion

On the surface, it seems many users accept responsibility for using SNS websites, and there are many options to handle problems related to information privacy abuse and online stalking. However, as described in this chapter, there remain concerns over the use of personal information and the management of digital identity on SNS. Updating personally identifiable information on SNS websites, our digital identity is shaped not only by ourselves, but also by other users and the architecture of systems. Bearing in mind that SNS is an ICT-based dataveillance system which is intended for profit, it is inappropriate to assume that the users have complete control over the construction of their digital identity. Self-responsibility for the development of one's digital identity has its limits.

Many SNS users may not consider such kinds of issues seriously and instead, actively incorporate information sharing on SNS websites that may be considered critical for advancing democratic principles. On the other hand, it is doubtful that the situation discussed here can really make us more autonomous entities on the SNS websites, because the construction of our digital identify cannot be controlled by ourselves completely. Individual autonomy is one of the most important values with respect to the protection of information privacy. The use of SNS may obscure this value.

At first it would appear that the spread of the value that "information sharing is a virtue" promotes the diffusion of hoarded information. However, nobody understands whether the mirror world on the SNS website really is leading its users to the utopian world it seems to be promoting, and that there is also a difference in perception caused by socio-cultural background. For example, it seems Japanese people have tended to favor the use of SNS which adopt anonymous or pseudonym naming systems compared to those in Europe or the U.S.A. Further research on SNS issues from these perspectives is necessary.

Acknowledgments This study was supported by the Strategic research foundation at private universities and Research Grant-in-Aid for Young Scientists (B) 24730320, from MEXT (the Ministry of Education, Culture, Sports, Science and Technology), Japan.

References

Allen AL (2008) Dredging up the past: lifelogging, memory and surveillance. Univ Chicago Law Rev 75:47–74

Asai R (2009) Classic yet contemporary gender norm: is ICT an amplifier of gender bias? In: Proceedings of CEPE 2009, Corfu, Greece, pp 18–26

Asai R (2010) NEKAMA men living different lives on the internet. Int Rev Inf Ethics 13:12–19

Battelle J (2005) The search: how Google and its rivals rewrote the rules of business and transformed our culture. Portfolio, New York

Clark RA (1988) Information technology and dataveillance. Commun ACM 31(5):498–512

Dwyer C, Hiltz S, Passerini K (2007) Trust and privacy concern within social networking sites: a comparison of Facebook and Myspace. In: Proceedings of the thirteenth Americas conference on information systems, Keystone, Colorado doi = 10.1.1.148.9388

Garling C (2011) Didn't get that new job? You need a better Facebook score. Wired Enterprise. http://www.wired.com/wiredenterprise/2011/11/reppify-identified-facebook-linkedin/

Goldhaber MH (1997) The attention economy and the net. First Monday 2(4–7). http://firstmonday.org/htbin/cgiwrap/bin/ojs/index.php/fm/article/viewArticle/519/440

Kirkpatrick D (2010) The Facebook effect: the inside story of the company that is connecting the world. Simon & Schuster, New York

Lawler JP, Molluzzo JC (2007) A study of the perceptions of students on privacy and security on social networking sites (SNS) on the internet. J Inf Syst Appl Res 3(12). http://jisar.org/3/12/

Mobile Marketing Lab (2011) About 80 % of Facebook users reveal their real name in profile, in mixi and Twitter website, about 20 % of Users. http://mmd.up-date.ne.jp/news/detail.php?news_id=784

Murata K (ed) (2004) Information ethics: individuals and organisations in the internet era. Yuhikaku, Tokyo [in Japanese]

Negoro T (ed) (2006) mixi and Second Net Revolution. Toyokeizaishinposha, Tokyo [in Japanese]

Odin Lab (2010) This is me. http://stores.lulu.com/odinlab

O'Keeffe GS, Clarke-Pearson K (2011) The impact of social media on children, adolescents, and families. Pediatrics 127(4):800–804, http://pediatrics.aappublications.org/content/127/4/800. full.pdf + html

Orito Y (2011) The counter-control revolution: "silent control" of individuals through dataveillance systems. J Inf Commun Ethics Soc 9(1):5–19

Rogerson S (2002) Computers and society. In: Spier RE (ed) Science and technology ethics. Routledge, London, pp 159–179

Weekly Toyokeizai (2012) Facebook: real identify of mega network which disclose information of billion users, 3 Mar 2012, pp 42–84 [in Japanese]

Chapter 14
Information-Offering by Anonymous Users in a Japanese Human Flesh Search

Analysis of Motivation in a Social Media Platform

Hidenobu Sai

14.1 Introduction

The development and diffusion of information and communication technologies (ICT) has led to the development of various tools and services. Among them, the tools and services that are referred to as 'social media' are increasingly being used. Social media include blogs, Twitter, social network services (SNS), such as Facebook, image and video sharing websites, such as Flickr and YouTube, and question-and-answer sites, such as Yahoo! Answers and Quora.

In this study, social media are defined as services or platforms in which many people can participate by offering and exchanging information. One characteristic of this kind of social media is that users attach greater importance to the information offered by and shared with other users than to information from a service provider. Thus, to be functional and attractive, social media have to encourage and motivate users to offer information.

Within this context, the objective of this study is to examine motivations behind information-offering in social media. In this chapter, we will begin by summarizing several previous studies that have focused on the reasons why individuals use various kinds of social media and their motivations for offering information. Next, we will present existing research about the 'human flesh search', which is a specific phenomenon in social media, and analyze a Japanese case of a human flesh search from the aspect of differences in information behavior online and offline. Finally, we explore the motivations behind contributions to human flesh searches and countermeasures against human flesh searches from the perspective of expectancy theory of motivation and information prospectability.

H. Sai (✉)
Faculty of Law and Letters, Ehime University, Matsuyama, Ehime, Japan
e-mail: saihide@LL.ehime-u.ac.jp

S. Uesugi (ed.), *IT Enabled Services*,
DOI 10.1007/978-3-7091-1425-4_14, © Springer-Verlag Wien 2013

14.2 Related Works on Social Media

Kawaura et al. (1998) identified three main groups of motivation for beginning to keep an online diary: motivation to express one's own thoughts, an instrumental motivation (e.g., making it easier to update information, or generating more traffic on a website), and a 'conformable' motivation (i.e., users may start an online diary due to influences from others).

Miura (2005, 2007) analyzed blog authors' intentions to continue blog-writing and identified three kinds of benefits to blog authors: benefits to self, benefits to relationships with others, and benefits to information-handling skills (Miura 2007). Miura surveyed blog authors and divided them into two groups based on characteristics of their blogs. One group is the 'database oriented blogger', who emphasizes sharing information and knowledge with a community. The other is the 'diary oriented blogger', who values expressing him- or herself, and maintaining closer relationships with others. For both types of bloggers, interactions with others affect their intentions to continue blog-writing. Additionally, benefits to information-handling skills are important to database oriented bloggers, but not to diary oriented bloggers (Miura 2005).

Heckner et al. (2009) studied information behavior in social tagging systems. According to their study, there are two motivations for tagging: personal information management for one's own use, and resource-sharing with others. Although these two motivations are important, users of Flickr and YouTube tend to focus more on sharing information and the users of Delicious and Connotea have tendencies toward personal information management.

Research on Wikipedia users (Glott et al. 2010; Schmidt et al. 2010) has found that almost 31 % of users contribute to Wikipedia as an author, editor, or administrator. Based on self-reported data, the most common motivation behind contributing to Wikipedia is the user's feeling that "I like the idea of sharing knowledge and want to contribute to it." The second most common motivation is "I saw an error I wanted to fix." On the other hand, the most popular reason for not contributing is "I don't think I have enough information to contribute." The second most common reason is "I am happy just to read it; I don't need to write it." Most users who do not currently contribute stated that they would contribute if "I knew there were specific topic areas that needed my help."

14.3 Human Flesh Search

Human flesh search (HFS) is a literal translation of the Chinese phrase "RenRouSouSuo" (人肉搜索), which means "searching by human power." Generally, HFS refers to cooperation in order to find particular information by a large number of Internet users. In many cases, the target of HFS is a particular person who has committed an illegal or immoral act. For this reason, HFS is sometimes called an 'Internet man hunt' or a 'cyber witch hunt'.

Although Wang et al. (2009) did not provide the original definition of HFS, they suggested that HFS shared characteristics with crowd sourcing, social search, collective intelligence, and distributed problem solving. They also pointed out five characteristics of HFS: accessibility, popularization, centerlessness, information timeliness, and convergence.

Wang et al. (2010) defined HFS as "A mass campaign, which comes into vogue through the medium of internet, targeting at searching for the identity of a certain person or the truth about a certain event, whose data collection depends partially on the human force to filtering the information gained from the search engine, and partially on the anonymous or real-name information announcement." The HFS process occurs within a short time, and in most cases, no matter how ambiguous or clueless the problem is, finds a correct answer. Wang et al. (2010) identified two characteristics of HFS. One is the importance of involvement of strong offline elements (e.g., information acquisition through offline channels or some kind of offline activism). The other is that almost all HFS rely on voluntary crowd sourcing.

In this study, we focused on HFS as a phenomenon that happens on social media and as one of the usage processes of social media.

One reason why HFS can yield information from relatively few clues is the fact that many users are engaged in search activities toward the same target, as opposed to one individual. The effectiveness of HFS, however, does not only depend on the large number of individuals involved; it also involves more sources of information than one individual can access. In this way, HFS can be considered to use the 'wisdom of crowds'. Surowiecki (2004) identified four main elements needed to exert the power of such wisdom: diversity of opinion, independence, decentralization, and aggregation. Because HFS involves a large number of users who come together from all walks of life through Internet, it tends to involve high level of diversity and decentralization. It is also highly likely to involve users from a wide range of backgrounds who can gather adequate offline information that cannot be found through online searches. However, although HFS involves high levels of independence, due to the lack of collective decision making, in most cases no aggregation mechanism is available. Thus, while HFS can use the power of the 'wisdom of crowds', it cannot maximize the benefits of this wisdom.

14.4 Differences in Information Behavior Online and Offline

To understand the characteristics of HFS, it is useful to set out certain concepts about information behavior online and offline because HFS is performed through information behavior online and offline. First, we can distinguish between "substance" and "information". Substance is a concept about a physical entity that exists only in the real world (offline). In contrast, information is an aggregation of symbols indicating some specific meaning and it can exist online and/or offline according to the media on which the information is recorded.

Offline	Difference	Online
Constant	Continuity of Sending	Intermittent
Not Always	Recording	Always Recordable
Difficult	Transmit to a Remote Place	Easy
Difficult	Searching, Deleting, and Duplicating	Easy
Difficult	Using Different IDs	Easy

Fig. 14.1 Differences of information behavior between online and offline

There are some differences in information behavior online and offline (Fig. 14.1). Offline, substance is always generating and sending information as long as it exists in the real world although its information is not always recorded. In contrast, most information sending through online activities is performed intentionally, and it is possible to send information intermittently online. Recording online information or transmitting it to a remote place is easier than doing this offline. This is also true for searching, deleting, and duplicating information.

Next, identification information can be distinguished from general information. Information indicating a specific substance is called identification information (ID). If the substance in question is a man or woman, his or her name can frequently be used as ID. However, degrees of tying between ID and substance differ by what kind of ID is used. Although the tying is strong if a real name is used as ID, it is very weak if the ID is no-name, that is, anonymous. In cases where ID is a pseudonym, the degree of the tying between ID and substance is located somewhere in between real names and anonymous. In this study, we use the word "identified" as a status that is strongly tied between ID and substance or real name.

When an individual wants to send some information, the sender can use different IDs, according to his or her situation. In contrast to information sending online, in which the use of information channels is limited, there is a tendency for information sending offline by using different IDs to become difficult because several information channels can be used simultaneously.

14.5 Case Study: HFS in Japan

In this section, we describe a case that occurred in Japan in December, 2010. Figure 14.2 illustrates timelines and platforms related to this case.

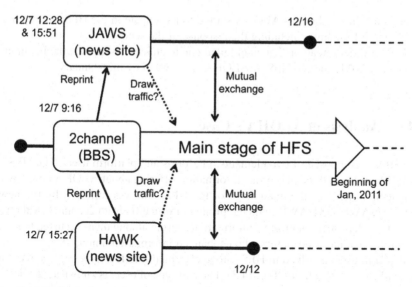

Fig. 14.2 Chart of ALOHA HFS case

On the evening of December 6, ALOHA (tentative name), a junior high school student, tweeted that he had obtained an illegal copy of a game just after its release. The next morning (December 7), ALOHA tweeted that he could not boot the file and asked how to boot the file at a popular Japanese question and answer website. In a short time, an interested individual found ALOHA's tweet and posted a thread about ALOHA on 2channel, Japan's largest BBS complex which has an enormous number of users. At this point, the HFS began.

Several hours later, some platforms reported ALOHA on the basis of ALOHA's thread on 2channel. Around noon, a news website JAWS (tentative name) reported ALOHA with screenshots of his tweets and his blog. After that, another news website HAWK (tentative name) reported ALOHA's story and then a comment on HAWK provided ambiguous information about ALOHA's place of residence.

On 2channel, the main stage of ALOHA's HFS, the HFS made progress in a short period of time. On the day the HFS started, HFS participants found ALOHA's other posts on a video sharing site and identified images of ALOHA. Additionally, they extrapolated information about ALOHA's school. However, they could not identify information about his school until next morning. Later, HFS participants extrapolated information about which junior high school ALOHA attended, based on the content of his blog and tweets. They shared a large amount of credible information but no identified information about ALOHA was available until the early morning of December 8.

On December 11, an individual went to a location where ALOHA took after-school lessons to investigate ALOHA, and another individual made contact with acquaintances of ALOHA using SNS. As a result, facts about ALOHA, including his real name, were identified on December 12. Then, students who attended the same junior high school as ALOHA started to participate in the HFS and reported additional information, including the response of the school.

The ALOHA thread on 2channel failed due to attack by scripts at the beginning of January 2011, and the HFS of ALOHA was settled by mid-January.

14.6   Analysis of ALOHA's Case

The HFS is composed of two elements: platforms and platform users. The HFS of ALOHA involved three platforms. Information-offering about ALOHA began with 2channel, which was the main stage of the HFS, and was followed by the news websites JAWS and HAWK, which reprinted information from 2channel. Although these are independent platforms, information sharing among them occurred to some extent through the contributors who exchanged information among them.

Platform users involved in HFS can be classified into four types on the basis of 'search' and 'post' activity (Fig. 14.3). The first type of users is known as 'Lurkers', who do not search or post anything. Lurkers are users who only read. The second type of users is 'Solitaries'. They search by themselves but do not post results. Solitaries are searching for self-satisfaction. The third type of users is 'Galleryites', who do not search but do post responses to HFS activity. In contrast, the fourth type of users, called 'Contributors' in this study, cooperate with searches and offer information to others. Contributors can be classified according to their main source of information: 'online' contributors and 'offline' contributors. Contributors can also be divided into two other groups: 'active' contributors who search and offer information continuously, and 'temporal' contributors who cooperate with HFS transiently. Thus, there are four types of contributors.

It is difficult to specify the contributors in the ALOHA case because almost all contributors posted anonymously to the HFS platforms. One exception was the thread about ALOHA on 2channel, in which contributors' posts could be identified to some degree because each post was displayed with a daily unique ID. Active contributors who were identifiable on 2channel included MANTIS (who investigated ALOHA's after-school lesson), HUNTER (who reached out to acquaintances of ALOHA), and MOLE (one of ALOHA's school mates) (all names are tentative). Of these contributors, HUNTER can be classified as an online-active contributor (On-A) and both MANTIS and MOLE as offline-Active contributors (Off-A). Figure 14.4 shows the main contributors in ALOHA's case.

Temporal contributors can similarly be classified into two groups: online-temporal contributors (On-T) offer information found in online searches or import information from other platforms, and offline-temporal contributors (Off-T) offer offline information, which is important for filtering and validating ambiguous information.

Generally, the process of HFS can be illustrated as follows (Fig. 14.5). First, information as a beginning (such as information about someone's illegal or immoral act) is offered on a platform. Of the participants who are interested in

Fig. 14.3 Types of users on HFS

Types of User		SEARCH?	
		NO	YES
P O S T ?	N O	Lurkers Reading only	Solitaries Self-satisfaction
	Y E S	Galleryites Responding but not searching	Contributors Searching and offering

Fig. 14.4 Main contributors in ALOHA's case

Types of Contributor		Source of Information	
		Online	Offline
Continuity of Contribution	Active	On-A HUNTER	Off-A MANTIS MOLE
	Temporal	On-T	Off-T

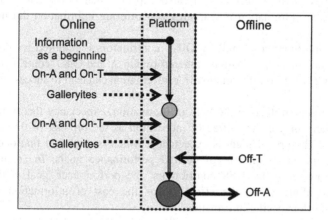

Fig. 14.5 General process of HFS

this information, On-A and On-T contributors start searching for online information and offering the results on the platform, although some galleryites simply encourage contributors and other galleryites post to stop contributors' actions. However, some information gathered and offered by On-A and On-T contributors may be ambiguous or incorrect. Then, Off-T contributors help to validate this information by using offline information. When a certain amount of credible information is gathered, Off-A contributors may begin to investigate or protest near the HFS target. On-A contributors sometimes request that Off-A contributors participate in these actions. Through all of these processes, On-A contributors search and offer online information and organize the information gathered, organize online protests, and provide direction for Off-A contributors. On-A contributors also tend to act as leaders in the HFS process.

14.7 Motivation of HFS Contributors

Pan (2010) argued that participation in HFS was motivated by two factors: acquisition of virtual currency and contribution to justice. Wang et al. (2010), however, found that fewer than 5 % of all HFS began with some kind of monetary reward. In the case of ALOHA, no contributors acquired real or virtual currency by means of the HFS. Because this particular HFS began with an accusation of illegal activity, it can be considered that one of the main motivations was a contribution to justice. Also, conformity to mood, which is created by galleryites and other contributors, played an important role. Additionally, benefits to information-handling skills probably had an impact on information-offering by On-A and On-T contributors as well as database-oriented bloggers. In contrast, Off-A and Off-T contributors might have been motivated to offer information by the awareness that the information was desired by someone. It is also expected that Off-T contributors only offered offline information if searching and offering information did not have a high cost.

Offline information offered by Off-T contributors is required to identify and validate ambiguous information gathered by On-A and On-T contributors. Thus, encouraging Off-T contributors to offer information can lead to success in the HFS process.

Of the research done in the field of motivation, expectancy theory provides a good explanation for the degree of motivation as the product of three elements (Fig. 14.6). The first element is expectancy, which is the belief that one's efforts will result in the attainment of desired performance goals. In an analysis of information-offering by HFS contributors, the performance goal is the act of information-offering and the effort means the cost of information behaviors required to offer specific information.

The second element is instrumentality, which is the belief that a person will receive a reward if the performance expectation is met. Many HFS contributors acquire only gratitude from others because very few HFS bring monetary rewards to participants. If HFS contributors are aware that information is desired by someone, they may expect gratitude from others for offering that information. Improving information-handling skills through HFS can also serve as a reward.

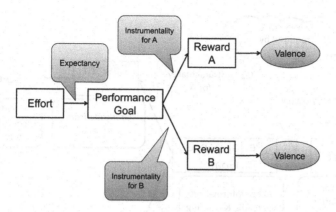

Fig. 14.6 Three elements of expectancy theory

The third element is valence, which is a positive or negative value an individual places on a reward. Regarding valence, there can be two types of rewards and valences for information-offering. One reward is gratitude from others, and the other is improvement in information-handling skills and the resulting feeling of achievement. The former has a high valence for offline contributors, and the latter has a tendency to have a strong valence for online contributors.

Instrumentality for gratitude tends to be increased when contributors are aware that specific information is required by others. On the other hand, instrumentality for improving information-handling skills can be subject to contributors' individual attributes. Thus, the degree of instrumentality depends on which kind of contributor will offer information (Fig. 14.7). However, even if instrumentality and valence are of a high degree for a contributor, if the contributor's expectancy is very low, information-offering will not be undertaken. This indicates that expectancy is the most important element relating to whether information is offered.

14.8 Cost of Information Behavior and Information Prospectability

We consider expectancy of information-offering to be affected by the costs of information behavior and information prospectability.

Contributors' information behavior is divided into three acts in this study (Fig. 14.8). The first is discerning desired information, or scanning gathered information. By this act, contributors can understand what types of information are needed. The second is additional searching. The third is offering information. Each act has its own cost and contributors estimate the cost to them. They decide whether to act by comparing each estimated cost and valence acquired in performing a particular information behavior. If the information required for the HFS process is specific, and a participant is convinced that he or she already has that

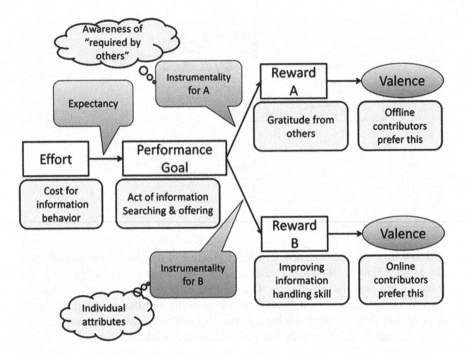

Fig. 14.7 Analyzing contributors' motivation by expectancy theory

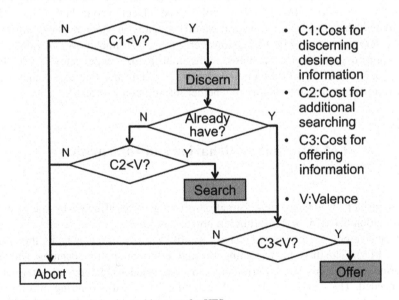

Fig. 14.8 Information behavior and its costs for HFS

information, the cost of information behavior for the HFS is only the cost of offering the information. To discern what kind of information is desired, a participant must scan information that has already been gathered on the HFS platforms. Thus, the participant has to pay time and cognitive costs. Additionally, if the participant does not have the desired information or cannot be convinced of its accuracy, he or she may not offer information or do any additional searching.

Thus, a prospect of obtaining information affects a participant's motivation or commitment to the HFS. In this regard, this study defines 'information prospectability' as a factor affecting the decision whether to offer information. Information prospectability is a subjective expectancy about the cost of information behavior required to obtain information about a target. The lower the cost of information behavior is required, the better information prospectability becomes. Thus, if information prospectability is high, in accordance with the expectancy theory of motivation, expectancy becomes higher.

Information prospectability depends on both individual attributes and the available information. The individual attributes include information-handling skills and self-efficacy. Available information means both the quality and quantity of information. Available information provides clues for additional searching and identifying ambiguous information. Excessive information, however, may increase the cost of scanning information. Moreover, information without coherency or context may become 'noise' and worsen information prospectability. Thus, ensuring high information prospectability requires not only increasing the amount of available information but also organizing and directing information. In HFS, these functions are provided primarily by On-A contributors.

14.9 Online and Offline Loops in HFS

In the HFS process, there are two 'loops', the online and offline loops (Fig. 14.9). Both loops start at with a limited offering of information by a contributor. In the online loop, if the cost for additional searching is estimated to be relatively small, online contributors search online and offer more information including noise. An increase of available information makes the cost for additional searching smaller. However, at the same time, the increase in noise makes the cost for discerning information higher and it decreases participants' motivation. According to the balance of the amount of information and noise, further online search may not occur.

The offline loop is the same as the online loop regarding estimating costs and offering information, although offline searches can reduce noise. Because of this, a combination of online and offline loops forms a positive feedback loop and it can provide more powerful searching than each loop alone. Nevertheless, for offline contributors, the cost of information behavior tends to be higher than for online contributors.

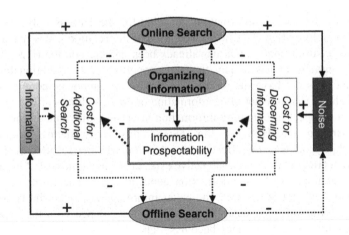

Fig. 14.9 Online and offline loops in HFS

Organizing information by On-A contributors makes information prospectability higher and the cost of information behavior smaller. That is one reason why On-A contributors enable more people to contribute to the HFS and accelerate online and offline loops. In HFS, repetition of these loops will improve the quality and quantity of information.

14.10 Conclusions and Implications

In this study, we examined the motivations behind information-offering on social media platforms, specifically using a case of HFS in Japan. HFS is a phenomenon in which a large, indefinite number of participants cooperate on a social media platform to find information about a target.

Contributors to HFSs very rarely acquire any monetary reward. Nevertheless, they still offer information. It is considered that the motivations of online contributors stem from benefits to their information-handling skills, although the motivation of offline contributors, especially Off-T contributors, is likely to be more related to the awareness that someone desires the information in question. In contrast, non-contributors do not offer information, even if they are aware of it.

We identified a new factor, 'information prospectability', as one that affects a participant's decision about whether to offer information. Information prospectability is subjective expectancy concerning the cost of information behavior required to obtain information about a target.

To improve information prospectability in HFS, On-A contributors play an important role by providing organizing and directing functions. Additionally, these functions can increase the number of participants and contributors in the midstream of HFS processes.

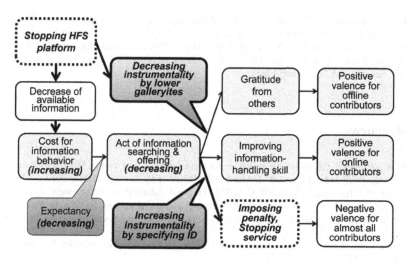

Fig. 14.10 Countermeasures against HFS

In this study, we clarified the HFS process using the expectancy theory of motivation and information prospectability. Further empirical research based on other case studies is required.

Finally, there are two implications arising from this study. One concerns encouraging users of social media to offer information. There are several approaches that can be used to encourage users. One involves offering rewards that have positive valences to users and increasing the instrumentality of rewards. In HFSs, galleryites' posts serve this function. Unlike HFS, in which information-offering is done by anonymous users in most cases, if users offering information contributed with specific ID, it would mean that previous records of contributions they had made could be shown and it can be some kind of rewards. Another approach to encouraging information-offering is decreasing the required cost for the information behavior. Although most of the cost depends on users' attributes, the cost for discerning desired information could be decreased by organizing information which is typically done by On-A contributors in HFS.

This study also indicates countermeasures against HFSs invading someone's privacy. These are also based on manipulation of the costs of information behavior, valence, and instrumentality (Fig. 14.10). Stopping the HFS platform brings about increases in information-discerning costs thorough being unable to use a place where information about the target of HFS is accumulated. Because stopping the HFS platform means stopping the 'theater' where many galleryites concentrate, it brings about decreases in positive valence and instrumentality. It is also effective to increase negative valences by imposing penalties for contributors. The development of laws against invading privacy would make negative valences and instrumentalities clear. Also regulation by a platform provider, based on terms of use, can be effective. Stopping the service for users involved in HFSs acts as a negative valence. Additionally, specifying users' IDs brings about an increase in instrumentality whether its valences are positive or negative.

Acknowledgments This chapter is based on the authors' presentation, "Information Offering by Anonymous in Japanese Human Flesh Search" at the 11th Annual International Symposium on Applications and the Internet, 2011 (with Yohko Orito). This study was supported by a Research Grant-in-Aid for Young Scientists (B) 23730354 from MEXT (Ministry of Education, Culture, Sports, Science and Technology) Japan.

References

Glott R, Schmidt P, Ghosh R (2010) Wikipedia survey overview of results. http://wikipediastudy. org/docs/Wikipedia_Overview_15March2010-FINAL.pdf. Accessed 1 Mar 2012

Heckner M, Heilemann M, Wolff C (2009) Personal information management vs. resource sharing: towards a model of information behaviour in social tagging systems. In: Proceedings of the third international AAAI conference on weblogs and social media (ICWSM 09), pp 42–49

Kawaura Y, Kawakami Y, Yamashita K (1998) Keeping a diary in cyberspace. Jpn Psychol Res 40 (4):234–245

Miura A (2005) Present and future of weblog. In: Yamashita K, Kawakami Y, Kawaura Y, Miura A (eds) The psychology of the weblog. NTT Publishing, Tokyo, pp 101–137 [in Japanese]

Miura A (2007) Psychological and social influences on blog writing: an online survey of blog authors in Japan. J Comput-Mediat Commun 12:1452–1471

Pan X (2010) Hunt by the crowd: an exploratory qualitative analysis on cyber surveillance in China. Global Media J 9(16):1–19

Schmidt P, Glott R, Ghosh R (2010) Analysis of Wikipedia survey data topic: reason for non-contribution.   http://wikipediastudy.org/docs/Wikipedia_NonContributors_15March2010-FINAL.pdf. Accessed 1 Mar 2012

Surowiecki J (2004) The wisdom of crowds: why the many are smarter than the few and how collective wisdom shapes business, economies, societies and nations. Doubleday, New York

Wang B, Hou B, Yao Y, Yan L (2009) Human flesh search model incorporating network expansion and GOSSIP with feedback. In: DS-RT'09 proceedings of the 2009 13th IEEE/ACM international symposium on distributed simulation and real time applications, pp 82–88

Wang E, Zeng D, Hendler JA, Zhang Q, Feng Z, Gao Y, Wang H, Lai G (2010) A study of the human flesh search engine: crowd-powered expansion of online knowledge. IEEE Comput 43 (8):45–53

Index

S. Uesugi (ed.), *IT Enabled Services*,
DOI 10.1007/978-3-7091-1425-4, © Springer-Verlag Wien 2013

Printed in the United States
By Bookmasters